D0408884

THEOLOGY AND SCIENCE AT THE FRONTIERS OF KNOWLEDGE

NUMBER TEN

CREATION AND SCIENTIFIC EXPLANATION

THEOLOGY AND SCIENCE AT THE FRONTIERS OF KNOWLEDGE

1: T. F. Torrance, *Reality and Scientific Theology.*
2: H. B. Nebelsick, *Circles of God, Theology and Science from the Greeks to Copernicus.*
3: Iain Paul, *Science and Theology in Einstein's Perspective.*
4: Alexander Thomson, *Tradition and Authority in Science and Theology.*
5: R. G. Mitchell, *Einstein and Christ, A New Approach to the Defence of the Christian Religion.*
6: W. G. Pollard, *Transcendence and Providence, Reflections of a Physicist and Priest.*
7: Victor Fiddes, *Science and the Gospel.*
8: Carver Yu, *Being and Relation: A Theological Critique of Western Dualism and Individualism.*
9: John Puddefoot, *Logic and Affirmation — Perspectives in Mathematics and Theology.*
10: Walter Carvin, *Creation and Scientific Explanation.*

THEOLOGY AND SCIENCE AT THE FRONTIERS OF KNOWLEDGE

GENERAL EDITOR – T. F. TORRANCE

CREATION AND SCIENTIFIC EXPLANATION

W. P. CARVIN

SCOTTISH ACADEMIC PRESS
EDINBURGH
1988

Published in association with the
Center of Theological Inquiry
and
The Templeton Foundation Inc.
by
SCOTTISH ACADEMIC PRESS
33 Montgomery Street, Edinburgh EH7 5JX

First published 1988

ISBN 0 7073 0487 3

British Library Cataloguing in Publication Data

Carvin, W. P.
Creation and scientific explanation. — (Theology and science at the frontiers of knowledge; 10).
1. Cosmology 2. Religion and Science — 1946–
I. Title II. Series
261.5′1 BD511

ISBN 0-7073-0487-3

Typeset by Pindar (Scotland) Ltd
Printed by H. Charlesworth & Co. Ltd., Huddersfield

CONTENTS

GENERAL FOREWORD

A VAST shift in the perspective of human knowledge is taking place, as a unified view of the one created world presses for realisation in our understanding. The destructive dualisms and abstractions which have disintegrated form and fragmented culture are being replaced by unitary approaches to reality in which thought and experience are wedded together in every field of scientific inquiry and in every area of human life and culture. There now opens up a dynamic, open-structured universe, in which the human spirit is being liberated from its captivity in closed deterministic systems of cause and effect, and a correspondingly free and open-structured society is struggling to emerge.

The universe that is steadily being disclosed to our various sciences is found to be characterised throughout time and space by an ascending gradient of meaning in richer and higher forms of order. Instead of levels of existence and reality being explained reductionistically from below in materialistic and mechanistic terms, the lower levels are found to be explained in terms of higher, invisible, intangible levels of reality. In this perspective the divisive splits become healed, constructive syntheses emerge, being and doing become conjoined, an integration of form takes place in the sciences and the arts, the natural and the spiritual dimensions overlap, while knowledge of God and of his creation go hand in hand and bear constructively on one another.

We must now reckon with a revolutionary change in the generation of fundamental ideas. Today it is no longer philosophy but the physical and natural sciences which set the pace in human culture through their astonishing revelation of the rational structures that pervade and underlie all created reality. At the same time, as our science presses its inquiries to the very boundaries of being, in macrophysical and microphysical dimensions alike, there is being brought to light a hidden traffic between theological and scientific ideas of the most far-reaching significance for both theology and science. It is in that situation where theology and science are found to have deep mutual relations, and increasingly cry out for each other, that our authors have been at work.

The different volumes in this series are intended to be geared into this fundamental change in the foundations of knowledge. They do not present "hack" accounts of scientific trends or theological fashions, but are intended to offer inter-disciplinary and creative interpretations which will themselves share in and carry forward the new synthesis transcending the gulf in popular understanding between faith and reason, religion and life, theology and science. Of special concern is the mutual modification and cross-fertilisation between natural and theological science, and the creative integration of all human thought and culture within the universe of space and time.

What is ultimately envisaged is a reconstruction of the very foundations of modern thought and culture, similar to that which took place in the early centuries of the Christian era, when the unitary outlook of Judaeo-Christian thought transformed that of the ancient world, and made possible the eventual rise of modern empirico-theoretic science. The various books in this series are written by scientists and by theologians, and by some who are both scientists and theologians. While they differ in training, outlook, religious persuasion, and nationality, they are all passionately committed to the struggle for a unified understanding of the one created universe and the healing of our split culture. Many difficult questions are explored and discussed, and the ground needs to be cleared of often deep-rooted misconceptions, but the results are designed to be presented without technical detail or complex argumentation, so that they can have their full measure of impact upon the contemporary world.

Dr Walter P. Carvin, the author of this work, has been interested in the relation between science and theology since his student days at Wheaton College Illinois and the University of Pennsylvania when he majored in physics and mathematics. He is also a graduate of Eastern Baptist Seminary, and of Princeton Theological Seminary where he took his doctorate under the direction of Professor Charles Gillispie of Princeton University. While he has been pastor at several Churches in New Jersey and New York states and is currently the minister of the First Baptist Church of Warren, Ohio, he also teaches religion and philosophy at Youngstown State University. He writes as both a mathematical physicist and a theologian who is deeply concerned with the relation between scientific explanation and the knowledge of faith. It is an analytical rather than a constructive work in which he finds the biblical concept of creation out of nothing, the cosmological argument for God and the science of cosmology to be closely related. Attention is focussed particularly upon the thought of Aquinas against the background of Aristotelian cosmology as understood in the Latin West, and upon the thought of Leibniz against the background of Cartesian mathematics and cosmology. In each case he finds that the concept of creation and the traditional cosmological argument for God must take their meaning and force from "the broadest scientific view of things". The implication for modern theology is that the doctrine of creation cannot be isolated from the cosmology of the day and that the dialogue between science and theology continues to be essential.

The Rev. Robert T. Walker of Edinburgh has very kindly joined with us in checking the proofs, for which we are most grateful.

Edinburgh, June, 1987

CHAPTER 1

CREATION AND COSMOLOGY

THOSE who oppose the attempt to relate religion to science often do so in the spirit with which the great formalist, Hilbert, sundered pure and applied mathematics. One day he was asked to substitute in a class at a nearby technical school for his colleague, Felix Klein, where Klein used his intuitive approach to mathematics to obtain practical applications. Hilbert said to the engineers, "I understand that there are those who feel there is some conflict between pure and applied mathematics. But this cannot be. They have nothing to do with one another." (*Sie haben nichts mit einander zu tun.*)

Obviously religion and science cannot be so unequivocally divided the one from the other. There must, for example, be *some* relation between modern science and the Biblical account of creation. But *what* is that relation? There is the nub of the problem.

One commonly accepted answer is that science deals with the observable and religion with the unobservable: one is "down to earth" and rooted in empiricism while the other is speculative and remote. But anyone who has studied modern physics will soon be disabused of the idea that science is "down to earth". The entities with which modern physics deal — the elementary particles — are so elusive, abstract and mathematical as to call for the highest training in abstract thought. On the other hand, the basic stuff of religion, that of religious experience, is immediate, emotional, and, for those who undergo it, very real indeed.

Another common answer is that religion deals with faith, science with reason and proof. Yet the history of science will reveal many methods of science, many changes not only of style but of substance. One need only compare pre-

and post-Copernican astronomy. There have been revolutionary changes in what, for science, "counts" as explanation. There is all the difference in the world between, for example, Aristotelian categories and those of Descartes. On the other hand, the man of faith will claim that he can give reasons for his faith, so that faith has its reasons and reason its seasons.

A better contrast between religion and science is in comparing them as alternate systems of explanation. If we ask a divine of the Presbyterian school why it is raining, we might conceivably receive the answer "because it is the will of God", and according to our own beliefs and anticipations accept that answer from him. But if we asked a meteorologist that question and he replied with the same answer, we should not be satisfied. We expect not exactly more from the scientist, but something else. We call it scientific explanation. We expect such explanation to be impersonal, objective, "value free" and enabling us to predict the future. We do not expect any reference to purpose or to God.

Now we know that the two explanations might well overlap at points. The clergyman might well make reference to a low pressure system; the scientist, if he is also a person of faith, might end by saying, "This is the entire system of weather as we have it now. Yet if you ask me why the system of meteorological laws is as it is, I must only answer that it is so because God wills it so." Yet the clergyman would have stepped out of role, and so would the scientist. What we hope for from the clergyman would be an answer which will provide some personal comfort, some spiritual sustenance to meet some personal problem (perhaps I had a picnic planned). What I expect from the scientist is something else.

But when we come to the concept of creation, we come to an area in which the relation between science and religion must be at its most intimate, if any such relation is indeed possible. For here the explanation of faith and that of science seem to be getting at the same thing: namely, why are things as they are; what is their origin? Here we push scientific explanation close to that which we expect from religion: an answer to the whole of things. And we ask of religion that its statements refer to the physical world, not to the soul and things invisible, but to the physical and tangible furniture of the universe. When God made the world he made, say the Scriptures, mountains and seas, and these are things which science studies. Thus here, if ever, the two systems of explanation must meet. Does science have anything to say about the origin of things? Does religion have anything helpful to say about the fabric of the universe?

Now in phrasing the question in this way, there are obviously some assumptions being made. The large one is that both science and religion do indeed refer to "things" that somehow exist in and of themselves and apart from our perception of them. I am assuming that there really are mountains and seas to be studied and that the scientists' study of them does not somehow so alter them that he can never say "the mountain" but only "my understanding of the mountain". It may seem that this is so obvious as to not need stating, but there are some philosophers of science who claim that one can never extract the scientist and his theories and ideas and models of scientific activity from that which he studies: that the scientist always muddies the waters in which he fishes. Now this may be a good thing to remind us of when we are looking at the esoteric particles of modern physics where one wonders if this jungle is not the invention of the physicist himself, but it seems so against common sense in dealing with large and immediately observable objects that only the most sophisticated of thinkers would consider it. I know that we have thought much of paradigms and models and the relativity of all knowledge since Thomas Kuhn wrote his *The Structure of Scientific Revolutions*, but is there really any astronomer today who doubts that the current view of planetary motion is better, i.e., more nearly the truth than the pre-Copernican view of epicycles? Can we really believe that phlogiston gets to the truth of things as well as modern atomic theory? Is there any doubt in the mind of the geologist or the paleontologist that he is studying an earth of multiple million years ago which was really here even though he was not, and whose form and nature is as he in fact describes it in his monographs?[1]

I assume much the same position in regard to religious affirmation. The first function of religion is not explanation but salvation; its primary tool is faith and not reason. Yet it does make affirmations about things as they are. The great affirmation is, of course, that God exists. In making this claim, the believer is not referring to himself but to a being which exists independently of himself. He is not referring simply to his own feelings. Granted that such statements are not religiously sufficient, yet they are meant to be cognitive affirmations. The same thing is true in affirmations about Jesus, about events of history related to the story of the Jewish Nation and the Christian Church, to the events of the resurrection. The believer is not simply stating things about the right way to live or about ethical beliefs; he is making statements about "what is". He says "I believe" but there is that which he *does* believe which is not identical with that act of belief, a "faith" that is not the same thing as his act of faith.[2]

Part of that "faith" is the affirmation of God as "Creator of heaven and earth". Here the believer is describing something about which science has something to say as well. Granted that he is not saying the same kind of thing about it that science does. He will not be making (unless he is a "creationist") statements about the age or composition of the world, yet he is referring to the same thing as science. Further, he is attempting a form of explanation of "heaven and earth". He is saying that the making of it by God in some way "explains" it. And, since science is in the explanation business, that explanation by religion and by science must come into some relationship to one another.

I understand by "creation" the classic Christian concept, that of creation *ex nihilo*; that is, as formulated by Augustine and Aquinas, that the world is utterly dependent upon God in each and all of its parts so that, apart from God, it would not "exist". This is an affirmation rooted in the Biblical faith and is a natural consequence of that faith. That God is "Lord" in all the full consequences of that word is part and parcel of Christian belief. Each and every thing of the world has its being from God. Were this not the case, God would be limited by something that was not of him, would not be truly free. This rules out any consideration of God which would make him in some way dependent upon the world, or which would limit him, making his role that of the guide and goal only, as in some form of process theology. It assumes that God is indeed transcendent to the world, and that the Christian affirmation of God as "Maker of heaven and earth" carries with it the implication that he is indeed thus transcendent. Any relation of the world to God which would make the world other than a free creation of God, which would make creation a necessary act of God or would put the world in its relation to God as other than something "made" by him will not be considered in this study. If, in the words of the Nicene Creed, the Son is "begotten, not made", the implication is that the world is made, and not begotten. For this classic understanding, the Son is to the Father as one begotten and thus necessary to the Father: the world is to the Father as a thing made and thus not necessary to the Father.

I wish to uphold these distinctions, not only because I believe them true and true to the Christian faith, but because it is within this nexus of belief that science as we know it in the western world was born and nurtured. When Aquinas, Descartes, and Leibniz wrestled with the deep problems of nature and of God, it is with some such assumption of the way of things that they began. It was with a world other than God, yet somehow related to him as a thing made to its maker, that early science did its investigation. It has been argued that it is because of this very relation, this otherness of the world that left it free for scientific investigation, that science arose on the climate of the West rather than the East.[3] That can be debated. But that this relation of Creator to creation is the one assumed by our scientific ancestors, there can be no doubt. It will be my contention that it is an understanding which will still carry freight, and it is the one by which we can best make sense of the relation of science and religion in our own time. I thus wish to maintain this classic concept of creation *ex nihilo* not only out of a sense of history but out of a sense of truth.

It is my contention that creation as thus classically defined in Christian theology is compatible with but not necessary to scientific explanation. The meteorologist, pushed for wider and wider patterns of explanation, may validly say in answer to the question, "But why are things like this?", "I know of no answer to that question. I do not know if there *is* an answer. Things may very well be the way they are because that is, in fact, the way they are. It is not my task to provide an explanation of the whole of things." Such an answer is a valid answer. Yet the answer of religion, "Things are the way they are because God made them that way", is itself perfectly compatible with it.

Yet to stop there is, from the viewpoint of religion, a risky thing. For the problem with science is that it never stops seeking explanations of everything under the sun, of man's own origin, of supposed miracles, of religious experience itself. Scientific explanation becomes ever wider and more comprehensive and self-sufficient until any other explanation seems, if compatible, irrelevant and superfluous. So the further, and more important, question will be: "Is there anything within science itself which indicates that it is open to that other system we call religion?" I would claim that when science has done its work and produced the widest system of explanation it can, namely a cosmology, there are still two things which ask for explanation which no cosmology can give: Why is there something and not nothing, and why is there *this* something, i.e., this ordered whole of which man himself is a part? It is in answer to these questions that the concept of creation bears relation to scientific explanation.

It is the science of cosmology which seeks the widest system of explanation of things and thus seems to come the closest to the concept of creation as religiously stated. It is assumed that cosmology is indeed a possible science: that one can indeed study "the whole" of things, can validly ask questions such as the age of the universe, if it is expanding indefinitely or oscillating in size, fill in details of cosmic evolution, ask concerning the total weight of the cosmos, investigate the largest scale of space and time. These are questions which are the bread and butter of cosmology. What, if any, relation have they to the concept of creation? It is in answer to this that we look at some historic statements of that relation, beginning with Genesis I and the Biblical account.

There seems little doubt that the account of creation as given in Genesis derives its picture of things from the Babylonians. For the Babylonians, the cosmos was a flat earth with the sky arched above it. In the sky were embedded the stars, sun and moon, rising and falling from East to West. The dome of the sky separated the earth beneath it from the watery expanse above. The origin of this is traced back to the battle between Marduk and the monster Tiamat. Tiamat has led a revolt against the gods, who dispatch Marduk as their champion. Marduk captures Tiamat in his net and then kills her with his arrows. He splits the carcass in half and establishes one-half as the heaven and the other as the earth, keeping the waters overhead at bay. He also establishes the stars. At the end of the account, man is created.

There is a definite parallel between Genesis and the Babylonian account. There is also a difference. The Babylonian account tells the story of the gods of which creation is but a part. For Genesis, there is no story of God, only of creation. Further, the account in Babylon features many gods, acting sometimes in concert, sometimes at cross purposes. The Genesis account has but one God who is in total control. For the Babylonian account, earth and sky are parts of the carcass of Tiamat. For Genesis, they are called into existence out of the chaos and are in no way part of God, but other than God. This ability to look at things as other than divine and God as other than things is distinctive in the Biblical account.

The key to the Genesis account is that of order. The primitive chaos gives way to cosmos. Light is divided from darkness, sea from land, earth from heaven, and each part of the cosmos has its distinctive place for living things: the fish in the sea, fowl in the air, man and beast upon land. What comes into being by God's fiat is a structured whole, and comes into being by step following step in orderly progression. Tiamat and Marduk have vanished: there remains only God, and his creative act.

The Genesis account is not to be symbolised or spiritualised. It is a straightforward and matter-of-fact statement of things as we know them, and how they came to be. It is spare, with no words wasted. It is not poetry in any subjective sense though the whole has majesty. It is ordered and causes the expectation of order, not of whim. That which is created, the natural order, becomes the enduring background of the actions of God in history. Mountain, sea and stars become steadfast signs of God's own steadfastness.

The Psalmist pictures this as he sings the praise of creation:

> Bless the Lord, my soul;
> O Lord my God, thou art very great,
> Clothed in majesty and splendor,
> And wrapped in a robe of light.
> Thou has spread out the heavens like a tent,
> And on the waters laid the beams of thy pavilion.
> Thou didst fix the earth on its foundation
> So that it can never be moved.
> The deep overspreads it like a cloak,
> And the waters lay above the mountains.
> At thy rebuke they fled,
> At the sound of thy thunder they rushed away,
> Flowing over the hills,
> Pouring down into the valleys
> To the place appointed for them.
> Thou didst fix a boundary which they might not pass;
> They shall not return to cover the earth.

Thou does make springs break out in the hills.
The wild beasts all drink from them.
The birds of the air nest on their banks
And sing among the leaves.
The trees of the Lord are green and leafy,
The cedars of Lebanon which he planted;
The birds build their nests in them,
The stork makes her home in their tops.
High hills are the haunt of the mountain goat
And boulders a refuge for the rock-badger.

Thou hast made the moon to measure the year
And taught the sun where to set.
When thou makest darkness and it is night,
All the beasts of the forest come forth.
When thou makest the sun rise, they slink away
And go to rest in their lairs:
But man comes out to his work
And to his labors until evening.
Countless are the things thou hast made, O Lord.
Thou hast made all by thy wisdom;
And the earth is full of thy creatures,
Beasts, great and small.
May the glory of the Lord stand forever
And may he rejoice in his works.[4]

The vision of the Psalmist and Genesis is the same, that of an ordered world. It is a vision based on observation, with the themes that appear throughout the Mediaeval period — the earth other than the heavens, that earth central, stable, unique and ordered by the wisdom of God and to the glory to God. The Genesis account is historically conditioned, to be sure, with its basic division of earth from heaven and its central place for earth. But what needs to be seen is not this specific order of parts but that the parts can be ordered. There is a rationality in this picture. There is a place for things, a structure penetrable by the mind. It may be a simple and homely structure, but structure it is. It is a view of things entire, a cosmos in which man has his place. It is a view adapted from the Babylonians but stripped of Babylonian mythology. As such, it is amenable to scientific explanation in a way in which the Babylonian was not.

For example, when a drought was ended in Babylon, it was because the gods had driven away the great bull whose hot breath dried up the land and a great eagle had come to shield the sun. Natural phenomena were explained in terms of the action of the gods. For the Bible, natural phenomena are also due to God — "The Lord makes the wind to rise." But the fact that the Biblical explanation is in terms of but one God permits an alliance with natural explanation. One expects more order and less whim from one God than from a council of gods who quarrel with one another.

Yet even if the concept of creation in Genesis permits an easier alliance with science than does the Babylonian, some cosmology is still needed if the concept of creation is to have any meaning. Note that it is not the particular cosmology which is important, but the fact that there *is* a cosmology. The Babylonian cosmology was arrived at by observation: by science, if you will. The simple observation that heaven is other than earth, that earth is stable and central comes because eyes are opened and look. That cosmology was then used in the Biblical narrative as an explication of the concept of creation. If creation as a religious concept is to have any meaning, it must have some cosmology by which it can be explicated. Thus, for one to say "I believe in God, the Father Almighty, Maker of heaven and earth", there must be some meaning to the words "heaven" and "earth". You say, "I believe God made this, and this — and everything" — throwing the arms wide to encompass earth and sky. And "everything" is not simply an enumeration. "Everything" refers to a cosmos. It may be as simple as earth beneath and sky above. It may be a relativistic finite but unbounded universe. But there is some total scheme of things to which one points or of which one speaks. This total scheme of things is cosmology. The scientist assumes that he can study a cosmos because it is one, and understandable. The man of faith needs that cosmos to explicate what he means by "creation". And, because he believes in one God who acts consistently, he expects that there will indeed be a cosmos, a structure penetrable by the mind.

Yet, while there is and must be some cosmology to which religion can point when it seeks to explicate the meaning of creation, that cosmology is not intrinsic to the Biblical concept of creation; the concept is not dependent upon it. Cosmology is not derived from faith. The cosmology of the Bible with heaven above and earth beneath, and a firmament above the heavens, is one which the Bible shares with the Babylonians. What is derived from faith is the fact that this cosmos is created by the free act of God. But the cosmology itself is derived from observation, from other cosmologies themselves derived from observation. The cosmology in the Bible is *shaped* by the Biblical faith. Its strong sense of order, the separation between creation and the Creator, all these are intrinsic to that faith. But the cosmology itself is not. The cosmology can change and faith need not end.

Biblical faith is not tied to a particular cosmology in the way in which it is tied to history. Were historical evidence to show that Jesus never lived, it is to be doubted that Christianity as now known could survive. Most certainly the content of faith would need radical revision. But the Biblical faith in God the Creator, "Maker of heaven and earth", a faith classically expressed in the doctrine of "creation *ex nihilo*" has already survived three transformations in cosmology and could conceivably survive another, in fact, may be in the process of doing so now. Much of this study will be engaged in sketching these changes and the thought of two theologians who, in the midst of change, sought to relate their faith to the new world picture of their day: who, on the one hand, accepted that picture, and, on the other hand, used it and deepened it to explicate the meaning of an enduring faith in God as Creator.

NOTES

1. Bas Von Fraassen, *The Scientific Image* (Oxford, 1980). Larry Lauden, *Progress and Its Problems: Toward a Theory of Scientific Growth* (Univ. of California, 1977).
2. Paul Tillich, *Systematic Theology* (Univ. of Chicago, 1967).
3. Alfred North Whitehead, *Science and the Modern World* (1925).
4. Psalm 104.

CHAPTER 2

THE WORLD OF ARISTOTLE

ARISTOTLE'S view of nature is one which has been described as organised common sense. He took the obvious facts of nature and made of them a coherent whole. The basic fact which Aristotle used to explicate nature is that of change, of becoming. He calls this 'motion' and it is a key to his thinking. It is a richer concept than our ordinary view of motion, with its simple idea of change of place. It is that for Aristotle, but much more. It means not simply a transfer of some object from one point to another, but change, becoming, always with a beginning point and an ending point. This fact of motion can be seen most clearly in the growth of a living thing from generation through maturity, but it can also be found in the burning of wood, the building of a house, the healing of a disease, the circling of the stars.

In her book *A Portrait of Aristotle*, Marjorie Grene argues that at the heart of Aristotle is a biological concept. Aristotle, she claims, was a biologist first and last and biology forms the key to his thought. Living things are discriminable, can be identified and specified, have a shape which can be recognised. Further, living things have within them a *telos* which lies at the base of that which is natural.

That is natural which happens "always, or for the most part" and is contrasted with what happens "accidentally, or by chance". Acorns obviously grow up to be oak trees. An acorn might be eaten by a pig, but that would be accidental to the acorn. According to its own intrinsic process of growth, it would, all things being equal, become an oak tree, not a cucumber vine. It moves by an internal principle to become its final and mature form, a form which can be identified and classified. All natural motion is this kind of motion. The full explanation of such motion gives us Aristotle's famous four causes: material, efficient, final, formal, with illustrations drawn from the growth of the embryo or the making of a statue.

Nature moves to an end, is teleological simply because that is nature's way. Each living thing has within it a principle, a *telos*, which brings it to its mature form, its end. Apart from this end, this mature form, there can be no way to understand what a thing is. One does not know what an acorn truly is until one sees that it is a young oak tree, and will in time become a mature oak tree. There is thus a logical sense in which, for Aristotle, the mature form always precedes the immature, the chicken always precedes the egg. The world is an interweaving net of process, of becoming and dying, a ceaseless hive of motion.

The motion of objects, or, as Aristotle puts it, "local motion", is part of this teleological web. Things fall to the earth because that is their place; fire flies upward because *its* place is in the heavens. The motion of each is explained by this seeking of natural place. The cosmos is one with a delineation of parts, a heavenly realm and an earthly. Space for Aristotle is not, as it has been for us since the scientific revolution of the seventeenth century, homogeneous and indifferent to the motion carried on within it. Space itself has a power. So Aristotle says in his *Physics*:

> The typical locomotions of the elementary natural bodies — namely fire, earth, and the like — show not only that place is something but that it exerts a certain influence. Each is carried to its own place, if it is not hindered, the one up, the other down. It is not every chance direction which is "up" but where fire and what is light are carried; similarly, too, "down" is not any chance direction but where what has weight and what is made of earth is carried — the implication being that these places do not differ merely in relative position, but also in possessing distinct potencies.[1]

Aristotle's concept of motion is tied to this concept of space. Space cannot exist without matter. There can be no vacuum. The *plenum* fills all.

Aristotle introduces the prime mover in book seven of the *Physics* where he notes that, since motion is eternal, and since all things that are in motion are moved by something else (even the acorn is enabled to become an oak tree, not only by its internal form, but because it is part of a host of other processes, as chemical change, the change of the seasons, etc.), there must be a first movement which is itself unmoved and eternal. Aristotle himself puts this succinctly,

> Since there must always be motion without intermission, and since there is nothing which moves itself, there must necessarily be something — that which first imparts motion, and this first movement must be unmoved.[2]

In the *Metaphysics* Aristotle expands this brief statement. He does so against a background of his analysis of motion in terms of potentiality and actuality. The acorn is a potential oak tree, the oak tree is an actual oak tree. All motion, or change, or becoming is a passage from potentiality to actuality. Water is potentially hot. But, in order for it to realise that potential, there must be something actually hot, i.e., a fire, which becomes in this case the efficient cause of boiling water.

That which is the end result of the process, that toward which the process is directed, is the form, or the substance, the "what it is" of the thing. It is substance in its primary form, and the most knowable of things. In Artistotelian terms, it is the *infima species* and corresponds to the end process of classification. We thus move, as we advance from potentiality to actuality, from the unformed to the formed, from the unknowable (prime matter) to the fully knowable (primary substance).

Again, Aristotle himself puts it best.

> Evidently even of the things which are thought to be substances most are only potencies — both the parts of animals (for none of them exists separately; and when they are separate then they too exist, all of them, merely as matter) and earth and fire and air: for none of them is a unity, but as it were a mere heap, till they are worked up and some unity is made out of them.[3]

Here Aristotle sees the elements of the world lying "as in a heap". As such they are not substance, they are pure potential, matter. It is only as they are made into a unity that they become substance. The essential thing is the form, the shape. As the tissues are material for the organs, so the organs are material for the living body, which is a more complex unity than the organs even as the organs are a more complex unity than the tissues.

The act of definition, so important for Aristotle, is akin to this, the genus standing to the matter as the differentia to the act, i.e., man as rational animal is better defined, more fully substance, than simply "animal". The final definition is the form and actuality of that particular thing, is in the fullest sense substance.

It is over the ceaseless change and becoming of such a world, with each individual thing reaching its end and culminating in its own individual substance that we find the prime mover as source. The prime mover is not the form of the whole in some Hegelian sense. Rather is it a separate and necessary substance which by the desire and imitation of itself causes the ceaseless coming-to-be from matter to form and the change from potency to actuality.

The prime mover is pure actuality, with no mixture of potentiality. It can have no potentiality, for that would mean it has the possibility of being other than it is.

> No eternal thing is potential. . . . That which may possibly not be is corruptible, either in the full sense or in the precise sense in which it is said that it may possibly not be. . . . All corruptible things are, then, actual. Nor can anything which is of necessity be

> potential; yet these things are primary, for if these were not, nothing would be.[5]

The world, for Aristotle, is full of change, of things moving, of things coming to be and ceasing to be. Each thing has the source of its motion not entirely in itself but in some other thing which moves it. For Aristotle, it is the stars which provide the efficient cause of this motion. In their ceaseless circling, they affect everything on earth. But the stars, despite their near perfection, typified by that most perfect of geometrical figures, the circle, are not a sure guarantee of motion, and motion must never cease. The stars must be moved by something which is pure actuality and which is itself not in motion, in short, by the prime mover.

The prime mover is thus a necessary being, i.e., it is not possible for it not to exist, for it is not possible for motion not to exist. By love of the prime mover, the stars are moved, and the stars in turn move everything else. The prime mover is the guarantor and explanation of motion in the world.

The prime mover is the explanation of motion. It will be well to pause and consider what such an "explanation" means for Aristotle, and do so in contrast to those against whom both he and Plato raised their systems, namely the ancient atomists. It is Plato who puts the issue most clearly in his famous last dialogue of Socrates, in that moment when death puts all philosophical issues in an ultimate light. He says that he was delighted to read in Anaxagoras that it was mind that had produced things, and expected to find an explanation of things in terms of mind. Instead,

> I discovered that the fellow made no use of mind and assigned it no causality for the order of the world, but adduced causes like air and aether and water and many other absurdities. It seemed to me as inconsistent as if someone were to say "The cause of everything Socrates does is mind — and then . . . said the reasons I am lying here now is that my body is composed of

> bones and sinews, and that the bones are rigid and separate — and that is the cause of my sitting here . . . and never troubled to mention the real reasons, which are that since Athens has thought it better to condemn me, therefore I for my part have thought it better to sit here. . . . But to call things like that causes is absurd. If it were said that without such bones and sinews and all the rest of them I should not be able to do what I think is right, it would be true. But to say that it is because of them that I do what I am doing and not through choice of what is best would be a very lax and inaccurate form of expression. Fancy being unable to distinguish between the cause of a thing and the condition without which it could not be a cause![6]

What Plato will explain in terms of mind, Aristotle will explain in terms of nature. Both would seek to distinguish "between the cause of a thing and the condition without which it could not be a cause". The views of such men as Empedocles that all can be explained in terms of the mingling of the elements of earth, air, fire, water, is rejected by them as simply inadequate. The atomists popularised by Lucretius who would explain all by atoms and the void, they would have also rejected.

For Aristotle, things are explained in terms of their ends. All of nature is thus explainable, definable. The natural "place" of fire calls it to that place, of heavy things call them to their place, to their "natural" place. The mind can comprehend nature because nature is comprehensible, circumscribed. Names can be given to things because the things are definite. The mind moves easily in such a world because it is already demarcated. The species already exist in forms ready to be understood. The cosmos is bounded. There is no void, no actual infinity, nothing that the mind cannot get a grip upon, no infinite chain leading off to further infinities that defy understanding. What is is a cosmos, an entirety complete in itself with no need to look beyond it. It is bounded on one side by the cycle of the elements and on the other, the cycle of the heavens. Below the elements is prime matter, an intellectual construct needed for change to be possible; above the skies the prime mover, an intellectual construct needed for that change to be eternal. The cosmos is closed and rational and the prime mover exists to ensure that it remains that way.

The physical shape of the cosmos reflects this metaphysical order. At the center is earth, the place in which it is natural for it to be. It is central, unique, stable. (Were space infinite, it could have no such central point and thus would not be discriminable for each portion of space would be like every other portion.) There is no question but that for Aristotle or for any educated and/or travelled Greek or Mediaeval, the earth is a sphere. On the earth there is a coming-to-be, and a ceasing-to-be, a generation and a corruption, a birth and a dying. Yet all that happens happens "always or for the most part", with each thing reproducing its own kind and each object released falling back to earth.

Around this central earth rotate the heavenly bodies, each embedded in its own crystalline sphere like so many bees in amber. The spheres "nest" in one another, the innermost being the moon, the outermost the stars. These spheres rotate endlessly and by that motion influence the motion of all things on earth. This motion had no beginning and never ceases.

Beyond this is, literally — nothing. All that *is* is in this cosmos. This is a hard idea for us, with our inherited concept of Newtonian space as something which exists in and of itself. But for Aristotle, the space of "place" is simply the boundary of a contained body. It is the "innermost motionless boundary of what contains".[7] The boundary is the "place" of what lies inside the boundary. The river bank is the "place" of the river which flows between it. The crystalline sphere of the stars is the "place" of all within it.

> If then a body has another body outside it and containing it, it is in place, and if not, not.[7]

And since the crystalline spheres have no other body

outside them, they are not "in" place; place is only "in" them.

Only in an ordered cosmos is there "place"; place arises with the ordering, not apart from it.

> When you have a homogeneous substance which is continuous the parts are potentially in place: when the parts are separated they are actually in place.

These laconic words are Aristotle's equivalent of Genesis 1, though for him, there is no "creation". For Aristotle, a homogeneous and undifferentiated mass, one with no structure or form, has no "place". But when once the elements have been structured, then place comes to be.

The world in Genesis 1 comes to be from a separation and ordering. In Aristotle, it is this which gives rise to the very "place" in which nature is able to proceed in its motion from generation to corruption and over which the stars preside. For both Genesis and Aristotle, the world is bounded, picturable, rational, a whole.

Thus from both Genesis and Aristotle, Aquinas inherited a cosmos usable for the task of a Christian theology of creation.

NOTES

1. Physics 208b8.
2. Physics 258b10.
3. Metaphysics 1040b.
4. *Ibid.*, 1050b.
5. On Generation and Corruption.
6. Phaedo.
7. Physics 212b5.

CHAPTER 3

THOMAS AQUINAS: FROM THE WORLD TO GOD

WITH Aquinas, we come to a massive and consistent attempt to relate the assertion of faith in God the Creator with a cosmos whose features are derived from science. The transition from the Babylonian cosmology to that of the Greeks took place with some protest in the Christian community. There were those who argued against the Greek, or Ptolemaic, system of astronomy on the grounds that Scripture clearly teaches that the earth is flat since there are references to its "corners". And then too, if the earth is round, where, it was argued, did Christ go when he "ascended" into heaven: which way was "up"? Fortunately, good sense prevailed and scholars accepted the fact of the earth as a sphere. There was still, however, and despite the change, a distinction as in Scripture between heaven and earth. And the earth remained a central and stable platform, a fit stage for the drama of redemption.

It was with the coming of the full Aristotelian corpus that the difficulty began, and the difficulty arose, not with the shape of the earth, but with its supposed age. Plato held to a beginning of the earth which squared well with the Christian belief in creation. But Aristotle, now that his works were available in full, clearly held that the earth was eternal. He further seemed to hesitate on the question of the immortality of the soul. For these reasons, his work became suspect, indeed, for a time, banned. Yet in the magnitude and scope of his thought, the work of Aristotle had no equal. He was, for Aquinas, "the philosopher". Aquinas felt that Aristotle's metaphysics provided an exemplary tool for the task of the theologian. He implicitly accepted the cosmology of Aristotle and used that Aristotelian and Greek model as the background of his attempted proof of God.

Aquinas accepted the metaphysics, the distinction between potentiality and actuality, the parallel distinction between matter and form. He also accepted the physics which was part and parcel of that metaphysics. The concept of motion which he used in his first "way" to God is Aristotelian. The cosmology of Aristotle, that is, the widest attempt at explanation within a scientific framework, is the cosmology of Aquinas.

Now there can be no question but that Aquinas was a Christian, and that he put the metaphysics of Aristotle to Christian use, in the process bringing to it a new depth. But as he developed his ways to God, he was content to use the Aristotelian concepts as Aristotle used them so that the God who is the result of that thinking is the God of Aristotle, the prime mover, and not yet God the Creator, Maker of heaven and earth. Aquinas starts with Aristotle, and it is only when he has led the reader deep into the *Summa Theologica*, namely even to the chapters on the creation, that he develops that concept of God which is uniquely Christian. That is not to say that Aquinas is a pagan in the first part of the *Summa* and a Christian in the second. He philosophises in the light of faith. He is a Christian thinker. But when he would establish the existence of God, he is content to use those guides that Aristotle placed in his hand and is content for the moment with the God to whom those guides will take him.

If we look at the first "way", that from motion, the Aristotelian categories with which Aquinas is working become clear. Aristotle, like Aquinas, is an empirical philosopher. He seeks to relate his metaphysics to obvious things, things to be seen by anyone with eyes in their head. One of these plain facts is that of motion, of change. Things move from one place to another. Things change from one state to another. The fire heats the wood and the wood is reduced to ashes. A man moves his head and the staff in his hand moves also. What is the explanation of this change, this motion?

The explanation, says Aquinas like Aristotle before him, is that things are not only as they now are but as it is possible for them to be. The wood is indeed wood but it is also potentially ashes. The fire changes the wood to ashes. Motion is a change from potentiality to actuality and, in order for such a change to take place, there must be something already actual to effect the change. Things act "only insofar as they are actual".

So for Aquinas and Aristotle, change, and motion as an illustration of change, are explained by a metaphysics of potentiality and actuality. Motion, for Aristotle, is the "act of a thing in potency insofar as it is in potency". A thing changes because there is something which already is in actuality which can act upon it. This is part of a cosmology which symbolises it. The stars are the realm of pure actuality, symbolised by the fact that the stars do not change, except in rotation. The earth is the realm of change, of generation and corruption, of things which come to be in one form and pass into another form. But the stars endure, and ensure that motion and change shall continue. Things fall to the earth because they have a potentiality to do so that is actualised as they reach their goal. Neither Aristotle nor Aquinas was thinking in terms of a "billiard ball" theory of motion, where to say that "everything that moves is caused by something else in motion" is to think of an endless chain of billiard balls stretching back into infinity, each one hitting the next forever. There would be no logical reason for there not to be an infinite chain of such billiard balls, and if God were an explanation for such motion, it would only be as one who pushes the cue stick in order to start the whole chain in motion. Aquinas, like Aristotle, is seeking for an explanation of motion, not simply as a change of place, but as a change from potentiality to actuality, and thus for neither would an explanation simply in terms of previous "motion" suffice.

The argument thus is that since motion is a change from

potency to actuality and must be caused by something already in motion, hence in actuality, there must be that which can move all else without being itself moved or we shall have an infinite chain of explanation and thus no explanation at all. There must be that which explains all other motion and is itself unmoved, i.e., is in pure actuality. This is, for Aristotle, the Prime Mover, and for Aquinas, God. The cogency of the argumentation is questionable for us but plausible, given the Aristotelian view of motion. The main point is that we are dealing with Aristotelian categories through and through.

The motion and change are teleological. It is change towards an "end", something which is not now but which can be and which is an intrinsic part of the thing changing. The paradigm, even if the examples given do not always express it, is biological. The tree is part of the acorn though there is only the acorn, even as the acorn is part of the tree even if there is now only the tree. Motion, change, is never simply what it is now but also what it yet shall be. A snapshot of the present plays false, only the moving picture which indicates the future contained in the present shows true.

It is *this* kind of motion which is explained by the Prime Mover. Always the explanation must match that which is seen as needing to *be* explained. In this case, what needs explanation is the plain fact of change as change *towards*, thus requiring not simply a previous motion of that instant but a motion which is actual and itself "towards" to explain it. It is this "hiddenness" of things — hidden perhaps at the moment but intrinsic and seen in the actuality — which makes a simple "one level" explanation inadequate. (We shall return to this theme when we look at Descartes.) An infinite chain of movements is plausible if one thinks of motion as we moderns have been taught to do since the seventeenth century: but not if one thinks of motion in this teleological framework. It is claimed that the first way to God is invalid because it is based on a false physics. It is true that the physics of local motion as expressed by Aristotle was found inadequate. But when it is seen that

that physics was a part of a teleological metaphysics, its cogency within that metaphysics becomes clearer. If the motion we seek to explain is that of Descartes, the first mover is indeed superfluous. But if we seek to explain teleological change, then the pure actuality of Aristotle, whom Aquinas names as God, makes sense.

With the first way as background, we can turn to the second. This is usually assumed to be an early form of Leibniz's "why is there something and not nothing?" argument, or argument from contingency, sometimes referred to as the cosmological proof. I would maintain that while cosmological, it has a teleological overtone to it and raises the question "why is there *this* something, namely an ordered universe, instead of something else, namely a disordered universe?"

It will be well to look at the text.

> The third way is taken from possibility and necessity, and runs thus. We find in nature things that are possible to be and not to be, since they are found to be generated and to be corrupted, and consequently, it is possible for them to be and not to be. But it is impossible for these always to exist, for that which can not-be at some time is not. Therefore, if everything can not-be, then at some time there was nothing in existence, because that which does not exist begins to exist only through something already existing. Therefore, if at one time nothing was in existence, it would have been impossible for anything to have begun to exist: and thus even now nothing would be in existence — which is absurd. Therefore, not all beings are merely possible, but there must exist something the existence of which is necessary. But every necessary thing either has its necessity caused by another, or not. Now it is impossible to go on to infinity in necessary things which have their necessity caused by another, as has been already proved in regard to efficient causes. Therefore we cannot but admit the existence of some being having of itself its own

> necessity, and not receiving it from another, but rather causing in others their necessity. This all men speak of as God.[1]

Now the problem with our reading of this text is that we usually come to it with the assumption that Aquinas is here speaking of God the Creator, who summons all things into being *ex nihilo* and poises them over nothingness. We thus read "ceasing to be" in the sense of "contingent" and "contingent" in the sense of "dependent upon" and "dependent upon" as implying an alternative of either being or not being. Existence is then with a possibility "not to be" in the radical sense of not being *at all* and this, in turn, implies God as Creator *ex nihilo*.

But is this what Aquinas intends? Note that Aquinas is here developing his proof from experience as befits an empirical philosopher in the style of Aristotle. It is a fact of our experience that things change, thus there must be something changeless (the proof from motion). It is a fact of our experience that A causes B, thus there must be something which causes and is not caused (the proof from cause). In the same way it is a fact of our experience that things come to be and cease to be. But in what sense do they cease to be? Not in an absolute sense. Not in disappearing into *nihil*. This desk "ceases to be", which means it is no longer a desk but a pile of ashes. My experience is never of anything disappearing into a vacuum, but of gradually changing into something relatively formless.

This is the process of "generation and corruption" to which Aquinas points as the starting point of his proof. He is here taking over the thought of Aristotle. Things come to be and cease to be, become formed and lose their form through the action of something which is already formed. The formed never comes from the formless. But the formed will, in the process of time, lose its form and return to the formless. Another word for the formless is "prime matter".

Any other interpretation of "coming-to-be and ceasing-to-be" for Aquinas runs into the difficulty that for

Aquinas, things never cease to be in any radical sense of *nihil*. Thus, in his article on "whether anything is annihilated" Aquinas says

> Now the nature of creatures shows that one of them is annihilated. For, either they are immaterial and therefore have no potentiality to not-being or they are material, and then they continue to exist, at least in matter, which is incorruptible, since it is the subject of generation and corruption.[2]

> All the creatures of God in some respects continue forever, at least as to matter, since what is created will never be annihilated, even though it be corruptible — for corruptible creatures endure forever as regards their matter, though they change as regards their substantial form.[3]

Form may come and go. Matter is forever. Granted that God could, if he wished, annihilate things, but in fact he does not. Therefore, since the proof is based on experience, it cannot be the experience of a radical ceasing-to-be but only of this relative ceasing-to-be, of generation and corruption as commonly observed.

Now look at the phrase, "everything can not-be". To what is Aquinas referring? Is it to individual existences taken separately, or to all of them together in an inclusive sense? Is he referring to individual existences of the cosmos or the cosmos itself as a whole? It seems that the latter might well be his meaning. What he might have in mind is not individual coming-to-be and passing-away within the cosmos, but that cosmos as itself a whole, in particular, an Aristotelian whole.

The argument is then this: that since the cosmos as a whole is something which can corrupt and lose its form, it would, given infinite time, do just that and, once lost, that form would never be regained. But since this has not happened, there must be something which retains its form and could not lose it, some "necessary being". Here, "necessary being" is one which is pure actuality, which, in Aristotle's cosmology, could be a star; in Aquinas', could be an angel. An angel is pure form. The difference between an angel and God is that an angel could be annihilated (it would indeed vanish *in nihil* since it contains no matter) and thus does not have "its necessity from itself", while God cannot cease to be, and thus has his "necessity from himself".

In this understanding of Aquinas, the world is not poised between God and the abyss of nothingness, but between pure formlessness (equivalent to Aristotle's prime matter) and necessary being on the other. Thus the question posed is not "why something and not nothing" but "why *this* something (the highly structured Aristotelian cosmos) and not something else, namely pure formlessness".

This has advantages over the usual interpretation of the third "way" to God. For one thing, it avoids the obvious objection which is raised against the proof "what about matter?" For even in Aquinas' time there were those who felt that atoms endured, and could easily counter the usual interpretation of "cease to exist" by stating that, for their philosophy, atoms did not cease to exist. The objection continues in our own day with those who note that while things may cease to exist in a material form, they continue to exist as energy. Phrased as a form of teleological argument, the third way is immune to this particular criticism. For Aquinas would agree that things do not cease to exist in a radical form, that "something" does indeed persist, namely prime matter.

Further, this interpretation makes better sense of the "necessary beings" Aquinas introduces in the second half of the proof. These beings are clearly either stars, souls or angels. They are beings of pure form. Because they have no matter, they cannot be subject to the usual law of dissolution, and thus endure. But also, because they could be annihilated by God, they do not have their "being from themselves". They are eternal, but not self-eternal. The parallel is with Aristotle's proof of the prime mover, where he first introduces the stars as actualities with no possibility of not being, with the one exception of their change of place. This bit of potentiality in them leads one to the pure actuality of the prime mover.

To embed the argument in its Aristotelian framework is not to destroy its effectiveness when it is seen as a teleological form of the cosmological argument. The cosmological argument moves from the existence of anything, and the possibility of its not being, to God. But the teleological argument moves from this world, this particular world, to God, asking not "why is there something and not nothing" but "why is there *this* something and not something else?" The nature of the "something" from which you start has something to say about the effectiveness of the argument. If the "something" from which you start is in a low state of order, a gas for example, there seems no reason for its existence. The question "why this something?" seems to have no pertinency. But if that from which you start is highly ordered — an Aristotelian cosmos, for example — then the question has point. And the more highly ordered the cosmos, the more pertinent the point. In this way the historical setting of the argument is an advantage rather than a disadvantage. After all, we must always start with what we have, even as Aquinas did.

This formulation of the argument also has the advantage of not relying on a concept of "nothing". When you ask "why is there something and not nothing?", the word "nothing" is suspect. We have no experience of "nothing", only of "somethings". We can, by an act of imagination, think away one thing after another until we are left with nothing at all. But to do so is to go beyond experience into the realm of thought where we begin to wonder if the question is in fact a valid one. But in the teleological argument, we ask "why is there this ordered world instead of a disordered one?" and both are part of our experience. We know what disorder is. We can well imagine it to be pervasive. We can easily imagine the gas of outer space to be all there is, or the world to have some other physical structure which would, for example, preclude life.

The point of all this is quite simple: namely that this third "way" to God in Aquinas begins with the world as Aquinas knew it. Given an Aristotelian cosmos, given the premiss of the primacy of actuality over potency, the argument makes sense. Remove the argument from that world and it does not. The modern critics of Aquinas, notably Kenny in his *The Five Ways*,[4] have seen clearly the Aristotelian cast of his thought and sought to show that the argument is unsound. By modern standards he is right. But what Kenny does not see is that it is this very historical conditioning of the "ways" to God that is their strength. The explanation that you seek of things depends upon that which you perceive as needing to be explained. The "way" to God from the world must always start with that world in which you are living. Aquinas does start here. The problem is, does he indeed get to God? To God the Creator of heaven and earth? And the answer must be that even in Aristotelian terms, he does not. Aristotle went from the world to the stars and paused there. He posited a prime mover, but the prime mover simply explained the continuance of motion. Aquinas gets to his "necessary being" with rather a cogent argument, but how you get beyond that to God seems dubious. We might understand the meaning of the sentence "necessary things have their necessity from themselves or from another" in the sense that an angel is a created thing but does Aquinas show that this is so? Why not, as with Aristotle, simply say that this is the way nature works, include the stars and prime mover as explanation of that working, and stop there? I think the answer is that these are arguments which are more than the sum of their parts. The "more" is the cosmos of which they are expressive. Aquinas wants to ask of this cosmos not simply why motion continues in it but why it exists. He does so because he already has in his mind the Christian way of looking at things: he knows that God created the world, that all this comes to be, in the most radical sense *ex nihilo*, from God. The fact of existence is one which does, in fact, call for an explanation. But it does so only from one who already sees it as a fact which *needs* explanation. For Aristotle, the thought that the cosmos might not even exist never even seems to arise, so he seeks no explanation for that existence.

Aquinas does not introduce a Christian concept of creation into his "ways" to God. The "ways" to God are ways to the God who orders, guides, but not to the God who creates, at least not in this early part of the *Summa*. Aquinas knows that the fact of order, of a cosmos, moves the mind from that cosmos to God. He also knows that this is not all. There is not only the fact of order but of existence to be explained. But to do that, one must start, not with the world, but with God.

NOTES

1. *Summa Theologica* q.2, a.3.
2. q.104, a.4.
3. q.65, a.1.
4. Anthony Kenny, *The Five Ways* (1980).

CHAPTER 4

THOMAS AQUINAS: FROM GOD TO THE WORLD

THE thesis has been that the proofs of God in Aquinas assume an Aristotelian cosmology: a cosmology based on experience and common sense, as well as reason. The thesis is that the third proof remains within the framework of this cosmology and that the contingency there expressed is based on the experience of the coming-to-be and ceasing-to-be of things, not in the radical sense of existence compared to nothing at all, but the everyday sense of existence in one form in comparison to another. The thesis is that this third "way" to God follows that Aristotelian pattern by moving, not from the world of becoming and corruption directly to God, but to the "eternal beings" who are stars/angels and who are themselves part of this same cosmos, as they are for Aristotle. In all of this Aquinas remains within an Aristotelian framework. The cosmos of his exegesis is one derived from Aristotle, that is, from reason and experience, without the intervention of any notion derived from revelation. The cosmology which he adopts has its own integrity entirely apart from faith.

It is otherwise when we come to the concept of God, as indeed we should expect it to be. For, while the concept of God is derived from experience through analysis by reason, the full nature of that concept is not one which reason can easily explicate. This is quickly apparent when you turn from the article on the existence of God to the article on the simplicity of God. God is simple, says Aquinas, because he is pure form and pure actuality. There is in him neither matter nor potentiality. There is so far no difference in this from Aristotle's prime mover. For Aristotle, the stars are indeed pure form, the actuality which insures the continuance of motion, since, as pure form, they cannot cease to be. Yet even the stars have an element of change in them: they revolve. It isn't much of a change, but it is some. So there must be something besides the stars, some pure actuality which has no potentiality for change at all.

Thus far, Aquinas follows Aristotle. But now there comes a difference. Not only, says Aquinas, is there a relation between potentiality/actuality: matter/form in the world which leads on to the stars and angels which are pure form and, to a prime mover who is pure actuality, there is another difference; and this one is drastic. There is a difference between the essence and the existence of things. Just what this exactly means is a subtle question, but it certainly is a difference never dreamed of by Aristotle, and a difference which draws a clear line between God and everything else. God alone has his essence equal to his existence so that he cannot ever not be. All other things, including the angels, have their essence different from their existence, so that it is possible for them *not* to be. And that "not to be" is the radical meaning of contingency, especially clear in the case of angels. If an angel ceases to be, it ceases to be in the total sense of the word, for there is no matter for an angel, so that, for the angel, it is not the case that the matter persists when the form changes. It is rather that *nothing* persists. If an angel would cease to be, there would be first an angel and then nothing at all.

It is possible for this to happen to an angel, says Aquinas. Indeed, it is possible for it to happen to everything that is not God. Thus, there is a clear line between God and everything else. God cannot *but* exist: everything else might well *not* exist.

Now this clear line between God and the world cannot be drawn within a strictly Aristotelian analysis. For Aristotle, the stars are pure form, but pure form which has in it a little potentiality due to change of rotation. The prime mover differs from the stars only in not having that potentiality. Likewise for Aquinas, the angels are pure form. But they have the potentiality for some change.

Their affections, for example, can change. Thus, they are not pure actuality, while God is. But *this* difference between an angel and God is only that one has a possibility of change and the other does not. From the world to the angels to God, we move with even steps. But the difference between God whose essence is his existence and an entire cosmos, including the angels, whose essence is not its existence, is not a step but a leap. One could say that it is a leap of faith. It is from this distinction between essence and existence in all else but God that Aquinas comes to the concept of creation.

To be other than your own existence means that your being is derived. And "derived" here is not something arrived at by a process of change or becoming, in the way of nature familiar to us. The being is derived from God, not simply generated by nature. Again, thinking about angels helps us understand what Aquinas means. An angel is its own essence: whatever there is about an angel is part of its form, since it contains no matter. But even an angel is not its own existence, i.e., it is possible for it not to be, even though by no process of nature could that happen. An angel is immortal only because God chooses for it to be immortal. It has its immortality not "on its own" but as a gift from God. But God is his own existence. There is no way for God to cease to be, for it is the essence of God to exist. Angels, while necessary beings to the extent that no process of nature could cause them to cease to be (and are thus sure guarantees, as with Aristotle, of the continuing order of the world), could yet cease to be if God so willed. But God is necessary being in the absolute sense: there is no process by which he could cease to be.

Thus, for Aquinas, all things under and in the heavens are in debt, not only for their form, but for their being. They are not only generated from one form to another, they are derived from God, and continue to be so.

> God is in all things: not, indeed, as part of their essence, nor as an accident, but as an agent is present to that upon which it works. Now since God is being

> itself by his own essence, created being must be his proper effect: just as to ignite is the proper cause of fire. But God causes this effect in things not only when they first begin to be but as long as they are preserved in being, as light is caused in the air by the sun. Therefore as long as a thing has being, God must be present to it, according to its mode of being. But being is innermost in each thing and most deeply inherent in all things . . . hence it must be that God is in all things, and most intimately.[1]

This is to see the world differently from Aristotle, to see it not according to form and matter, nor yet in terms of actuality and potentiality, but to see the world as "being". Each thing, form or angel, matter or star, while it may differ according to its potentiality for change, yet is alike in this: it has being. Thus to say that it "participates" in being is to say that it is contingent in a new and distinct sense. We say that it might not "be" and mean, not that its form may be other than it is, but that it might not "be" in *any* aspect. The alternative is not another form, or prime matter: the alternative is "nothing at all". And this is not an option which occurs by any process of nature. We do not experience things "ceasing to be" in this sense. Yet, says Aquinas, it is possible for things to cease to be in this sense. Did God choose for it to be so?

The correlate of this "ceasing to be" of things in this radical sense is the "coming to be" of things in the same radical sense. This is a coming-to-be not from previous matter, or from other forms. It implies nothing before it which conditions it. It is a coming to be *ex nihilo*, literally, "out of nothing". It is creation by God in its primal sense. God here is not an explanation of what is, but the origin of what is. God is not pure form or pure actuality arrived at from an experienced world but God is here pure being, not simply the one by whom things are explained (their motion, form, design) but the one from whom things are derived.

In the "proofs" we move from a world to God: in

creation we move from God to the world. The two ways do not imply each other. The way to God, for Aquinas, is based on experience and reason and has its own integrity. The way from God to the world also has its own integrity, and is based on faith. Let us examine it more closely.

First, we need to look at what Aristotle, and Aquinas, have to say about things infinite. Put simply, for Aristotle, the infinite does not exist. The reason for that is simple. The infinite, for Aristotle, means the indefinite. It is that which is vague, indeterminate, irrational. Only one kind of infinite is possible, and that is a potential infinity reached by the continual subdivision of a body. Since Aristotle denied the existence of indivisibles, i.e., atoms, any body can in principle be subdivided indefinitely. But, since this is a process which can continue, it will not reach an actual infinity. One can only say that if continued indefinitely, it would in time produce an infinite number. But apart from this rather special case, an infinite magnitude or multitude is impossible.

We catch the flavor of this reasoning as Aquinas considers the question whether an "actual existing infinite multitude" can exist (in distinction from one which is infinite only potentially).

> Now it is manifest that a natural body cannot be actually infinite. For every natural body has some determined substantial form. Therefore, since accidents follow upon the substantial form, it is necessary that determinate accidents should follow upon a determinate form: and among these accidents is quantity. So every natural body has a greater or smaller determinate quantity. Hence it is impossible for a natural body to be infinite.
>
> The same applies to a mathematical body. For if we imagine a mathematical body actually existing, we must imagine it under some form, because nothing is actual except by its form. Hence, since the form of quantity as such is figure, such a body must have some

> figure. It would therefore be finite, for figure is confined by a term of boundary.[2]

Everything which exists must exist as a *something*. G. K. Chesterton once said that the essence of a picture is its frame. Each existing thing must be determinate, thus finite, for the infinite is indeterminate. Given an Aristotelian world, the infinite is simply ruled out. Spiritual beings may be infinite in a relative sense as not contracted to a particular matter, but body as such is finite. The cosmos itself is finite.

This can also be seen by an analysis of motion.

> The same appears from motion, because every natural body has some natural motion, whereas an infinite body could not have any natural motion. It cannot have one in a straight line, because nothing moves naturally by such a movement unless it is out of its place: and this could not happen to an infinite body, for it would occupy every place, and thus every place would be indifferently its own place.[3]

An infinite body would occupy infinite space. But infinite space is indiscriminate space, space without a structure and without any privileged position. There could be no "up" or "down" in infinite space and thus no place for a body to go.

Thus this world could not be infinite. Such a world would be without form or determination. In a world in which form and the determinate are paramount, in short, in an Aristotelian cosmos, the world *must* be finite.

This is also implied in Aquinas' concept of creation. Creation is not the coming forth of any one particular being.

> A created thing is called created because it is being, not because it is *this* being, since creation is the emanation of all being from the universal being.[4]

One cannot go through the universe saying "this was created, and this, and this" in a process of enumeration. One says instead, "all being as *being* is created". This means that there must be come concept of "all being". In the cosmological proof, there is a problem called the problem of composition. It is that one cannot go from "A is a contingent and B is contingent and C is contingent" to the statement "A plus B plus C is contingent". That is why being able to look at the cosmos as a whole, a whole that is more than the sum of its parts, is important. The universe as a whole might not be as it is. So, just as we cannot move to that whole unless it exists as over and above the enumeration of its parts, so one cannot say "A is created and B is created" unless there is some concept of being as being, shared by A and B, in short, a cosmos. When we speak of creation, we speak of *all that is* as emanating from God. But this means that there is something which is "all".

Aquinas does not single out some particular thing or things and say "these are created: others are not". He says that *all* things are created. Creation is not some relation over and above some other relation more essential to it: it is the derivation of all being. Thus, "all being" is a thinkable concept. But if it were infinite, it could not, for Aquinas, be thought or have image; it would have no standing in the mind and the concept of creation itself would have no meaning, no instantiation.

Creation is *ex nihilo*. In distinction from generation and corruption, which assumes something from which the entity is generated, creation assumes nothing presupposed to it but God. It does not mean that God makes things "out of" nothing with that nothing considered as a sort of "thing" or possibility. Rather, it represents order, as "from morning comes midday" means "after morning comes midday". Yet even this is misleading. For Aquinas, there is no temporality intrinsic to creation. Aquinas says that the temporal language is simply "according to a mode of understanding". We can understand something as not existing and then existing only as if some change were going on, according to our usual experience: as, first the seed, then the tree. But creation is not according to our usual experience. It is not a change or motion. It is the derivation, not of a particular thing or a particular aspect of a thing but of being as being, of existence as compared to "nothing at all", with "nothing at all" as the alternative to "something". We state this in temporal terms of being "following" non-being only because this is how our imagination works. We think in space–time categories and find it difficult to make any picture in our minds that does not involve a change of being following non-being. But creation does not involve this "motion", so that

> In things made without motion, to become and to be already made are simultaneous. In these things what is being made, is. . . . Hence, since creation is without motion, a thing is being created and is created at the same time.[5]

Creation is a relation of things to God. It is a non-temporary relation of the dependence of all things upon God in the most radical of all fashions, namely for their very existence, their being.[6]

This might seem impossible. Surely creation is the bringing forth of this relation of radical dependence? If we could get behind the relation and see God bringing this relation into being, surely *this* would be creation? Not so for Thomas. In his laconic style,

> Passive creation is in the creature, and is a creature. Nor is there need of a further creation in its creation; because a relation is not referred to another relation but to itself.[7]

We cannot get behind the relationship of dependence of being. This relation *is* creation, and if we require another relationship to bring it about, then this will itself require another, and so to infinity. Somewhere the mind must rest, and it does so with the understanding of the dependence of being upon the source of being. Even had the world no first moment in time so that the world had been in existence since eternity it would still, says Aquinas, be true to say that God caused the world just as a man who had been standing in the sand for eternity could be said to be the cause of the imprint of his foot in the sand.

Since the concept of creation does not imply any temporal change, there is no way to say that creation necessarily implies a beginning of time. But neither does it rule out such a beginning of time. There is a difference here from Aristotle. For Aristotle, time is the measure of motion, and, since motion is eternal, time must be eternal as well. Time is tied into the world in such a way that it is impossible even to conceive one without the other. The circling and immutable heavens declare the unchanging nature of the world and thus the eternity of time. There is no way Aristotle could have a first moment of time.

It is different with Aquinas. He can have a "first moment" of time, for to him, God is Creator of the world and Creator of time together *with* the world.

When we are thinking about creation *ex nihilo*, we must be prepared to do without some of our common sense concepts. That concept of time is one of them. Whenever we are talking about change, we can speak of a time before that change. That is because we are dealing with a series of events, one event following the other. For Aristotle, the stars are a celestial clock to measure all this. But in creation, we are not dealing with change, with one thing following another. We are dealing with being as such: we have to do with the cosmos, all that is. We are saying that this cosmos depends upon God for its existence, does so constantly and universally and in each part. We are here at the boundary of our usual experience. Time is part of our ordinary experience, and something which we cannot project beyond that experience.

Time is part of the world and thus a created thing. So we cannot think of time as some independent and endless succession of moments of which we can say "There, that is when the world was created." There is no time outside of or before the world. Indeed, you cannot even use the word "before" here, since there is no way to measure it. If we conceive of some "time" before the world, it must be as we conceive of space "outside" the world. Had the world a first moment, it could be said that the world exists through all time in the sense that it has existed in all the time there is!

Imagine a time machine which could transport you backward in time, and you set the dial to year 0, month 0, day 0, minute 0, second 1. When you arrived at that first second, you would be no nearer to God than when you set out, nor would you be any "nearer" to creation. It would be impossible to go any farther backward to "observe" God creating the world, for time itself comes into being together *with* the world and at that very first instant, the world is utterly dependent for its being upon God — as it is now. There is no point outside the world and outside of God in which we can be spectators of the event of creation, for creation is not an event. Creation is a relation, and we are within it.

To say that the world has a "first moment" and to say that it does not is simply to say that in one case the dial of the time machine will have a first reading, and in the other case not. And it will be "first" only as regards the dial. There is no independent measure of time outside of the dial. At that "first" moment you cannot say "before this moment there is no world", for "before" will have no meaning. As you cannot travel to the boundary of the cosmos and say "outside of this there is nothing", for the world "outside" is meaningless, so the word "before" is meaningless when you would say "before" time.

The concept of creation does not depend upon a first moment of time. With or without such a first moment, the world has its being derived from God, and derived according to the will of God. If it is without a first moment, it is because God has eternally willed it to exist without a first moment. If it is with a first moment, it is again because God has eternally so willed.

The point of a first moment, then, says Aquinas, is

simply that it is fitting for God to thus "demonstrate his power".

> For the world leads more evidently to the knowledge of the divine creating power if it was not always than if it had always been, since everything which was not always obviously has a cause, while this is not so obvious of what always existed.[8]

It is simply easier to see that the world is created if it has a beginning in time than if it does not. The boundedness of the world, in time as well as in space, gives power to our concept of creation. We can say "this is what creation means. First there was no world and then there was one, and God is its Creator." The metaphysical subtlety of Aquinas' concept of creation is then instantiated in something at which the mind may look.

With the concept of creation as according to the will of God, with or without a first moment, we have come a long way from Aristotle. Aquinas uses Aristotle as he makes his "way" to God. He accepts Aristotle's cosmology. But when he comes to creation, he is talking about something which is not from the world of Aristotle, but from the world of faith. The concept of creation *ex nihilo* is not derived from Aristotelian cosmology. The world is dependent upon the will of God, not simply for its continuance in motion or its continuing generation, but for its being. Instead of the cosmos of Aristotle, there might be not some other variation or form, but simply nothing at all. And this cosmos, apart from the continued preservation by God, would cease to exist.

> As the production of a thing into being depends on the will of God, so likewise it depends on his will that things should be preserved in being: for he does not preserve them otherwise than by giving them being. Hence if he took away his action from them, all things would be reduced to nothing. . . . As it was in the Creator's power to produce them as things before they

> existed in themselves, so likewise it is in the Creator's power when they exist of themselves to bring them to nothing.[9]

The whole Artistotelian cosmology, beautifully articulate and in terms of which Aquinas develops his "way" to God and instantiates his concept of creation, simply might not *be*. It is contingent in the most radical of senses. In its place, there might be nothing at all.

It is a powerful concept, but not one Aristotle would have thought of. Aristotle, who abhorred the void, who demanded unceasing motion and postulated a prime mover to ensure the motion and the continued rationality of things, has his whole world suspended over the chasm of non-existence. But that chasm of contingency at its deepest would not be a chasm were there not a cosmos to suspend over it. Aquinas hangs the world from the will of God. But it was Aristotle who gave him the world.

It is interesting to note here the way in which Aquinas overcomes the problem posed by Aristotle concerning the eternity of the world. The problem was simply that Aristotle had asserted that the world had existed from eternity while, according to revelation, the world had a beginning. How was Aquinas then to use Aristotle when "the philosopher" was contrary to Scripture on so important a point? The solution is ingenious. It is that the question of the eternity of the world cannot, said Aquinas, be solved by science (and he meant the science of his own day). Reason itself cannot solve the question. Therefore, Aristotle has every right to adopt the view which in fact he did adopt. We know him to be wrong: the Bible tells us that the world had a beginning. But Aristotelian science maintains its integrity. It is simply that, from within that science, there was no way to decide the question. The question could only be decided from outside the science, namely by faith.

Aquinas here draws a line between conclusions to be reached by reason and those to be reached by faith, and, in effect, gives reason its charter to operate on its own side of the line. And the line is drawn, not with proof of God, but with the concept of creation. One can, says Aquinas, very well show the existence of God by reason. But a creation *ex nihilo* with its radical dependence of the world in its very existence from God, and the beginning of that dependence, can be known only by faith or, at the best, by reason tutored and led by faith.

Aquinas accepted the cosmos as it is known by reason and science. In a sense, he was doing what the author of Genesis did when he accepted the Babylonian cosmology. He accepts it as a means to explicate the very meaning of the term "creation". But what that cosmos is, what its structures are, its extent in space and in time, these are questions for science, not for theology. The question of the first moment is, for science, unanswerable; so for *this* particular scientific question, the answer of faith is supplied. In this case, what science could not answer, theology could. But since science had given its best answer, and a plausible one, one which is integral to it as a science, Aquinas received it with all respect but found it unconvincing.

As an illustration of how Aquinas felt about Aristotelian science, it is interesting to see how he interprets the first chapter of Genesis, always a crucial task for the theologian who seeks some relation with the science of his day. There is no difficulty with the first day. It is, properly, the work of creation, the other days being the work of separation (earth from water, waters above from waters below) and adornment (plants, animals, birds, fish). On the first day are created the angels, time, the "empyrean heavens" (i.e., the abode of the angels and future home of the blessed) and earth and sea. The heavens in some manner revolve, bearing with them on one side the matter of the sun (though the sun has not yet been formed from it), thus causing day and night. It is with the work of the second day, and especially with the making of the firmament, that Aquinas needs to wrestle hard with Scripture. There is no doubt in his mind but that Scripture speaks of the making of the world as Aristotle has described it. Yet the 'firmament' would seem to be the "heaven" of the first day.

In answering the problem, he gives a rule of Scripture interpretation derived from Augustine. First, to "hold the truth of Scripture without wavering". Second, that "since Scripture can be explained in a multiplicity of senses, one should adhere to a particular explanation only in such measure as to be ready to abandon it, if it be proved with certainty to be false; lest Scripture be exposed to the ridicule of unbelievers, and obstacles be placed to their believing".[10]

After giving a résumé of other opinions on this matter of the firmament, Aquinas advances his own, which is that by "firmament" is meant not the domain of the stars, but simply the "part of the atmosphere where the clouds are collected". It is this, not the "heavens", which is made on the second day.

The definition comes in handily when the problem is faced of the "waters above the firmament being divided from the waters below the firmament". For an Aristotelian cosmos, such a division of water above and water below is impossible. Water has its "place" on earth, not in the heavens. Yet this is what Scripture seems to say, as Aquinas admits.

> The text of Genesis, considered superficially, might lead to the adoption of a theory similar to that held by certain philosophers of antiquity, who taught that water was a body infinite in dimension, and the primary element of all bodies (Thales). Thus, in the words "darkness upon the face of the deep", the word "deep" might be taken to mean the infinite mass of water, understood as the principle of all other bodies. These philosophers also taught that not all corporeal things are confined beneath the heaven perceived by our senses, but that a body of water, infinite in extent, exists above that heaven . . . *as, however, this theory can be shown to be false by solid reasons, it cannot be held to be the sense of Scripture.*[11]

The true theory is, of course, that by "firmament" is

meant the part of the air in which clouds are suspended, and that this is what Moses *really* meant when he wrote in verse one that "darkness was upon the face of the deep". He meant by darkness, air. It is this air which is the firmament. He did not say so in plain words because, Genesis being addressed to ignorant people, they would not have understood him.

The interpretation is convoluted but the purpose is plain. Aquinas has no intention of deserting his cosmology simply because of a verse of Scripture. The plain sense of Scripture would imply water extending to infinity, and the firmament as making a division in this watery infinite. But this would make an infinite extension of place: it would mean heavens of essentially the same stuff as earth: it would mean the collapse of that tidy Aristotelian world which had afforded Aquinas both a way to God and an explication of the meaning of creation. He has no intention of deserting Aristotle for any theory shown false by "solid reasons". Cosmology has its own way of doing things, has its own integrity, and is not to be set aside by Scripture. Rather, Scripture must be interpreted to agree with what all agree is in fact the case. Aquinas had no more doubt that Aristotle was right in his description of the physical world than we doubt that Copernicus was right, and he would not hold Scripture up to ridicule by having it assert something which all knew to be wrong, any more than a churchman today would say that the sun revolved around the earth because Scripture spoke of it rising in the east.

Scripture, in short, is not the way in which one discovers the structure of the world. You discover the structure of things by using your eyes and your reason. Anyone with eyes in his head knew that the earth and the heavens were different. Anyone could see that the stars held their courses unwaveringly, while here on earth things changed, grew old and died. Anyone who observed could see that heavy things fell and sparks flew upwards to the stars. These were common things and, subject to the analysis of Aristotle, were part of a splendid scheme, an articulated whole, with a place for everything and everything in its place. It is just this structure of things which can lead you to God. It is just this structure of things which God created when he made heaven and earth. You could look at it: you could see it: you could think it through and it made sense. It is *this* world which God has created and of which Genesis spoke and, according to this world, Genesis itself must be interpreted.

Nothing could show more clearly the influence of Aristotle on Aquinas than this piece of exegesis. For here, in the very heartland of the faith, namely, in the exposition of Scripture, it is the Aristotelian model which decides the issue. When Aquinas would make his way from the world to God in his proofs, it is from Aristotle's world that he begins. When he would explicate the meaning of creation, it is Aristotle's cosmos which gives that explication. And indeed it must be so. The answer to the question "why something and not something else" must always begin with a particular something. The explication of the doctrine of creation must always be in terms of the world in which you live. It is *this*, this world, which God has created. Without some cosmos, there can be no meaning to the word "creation". Without a place to start, there can be no way to God. Aquinas took his place and his meaning from "the philosopher", and in that beautifully articulated cosmos found that which he could relate to his Christian faith, the world which God created.

NOTES

1. S. T. q.8, a.1.
2. S. T. q.7, a.3.
3. *Ibid.*
4. q.45, a.4.
5. q.45, a.2.
6. q.45, a.3.
7. *Ibid.*
8. q.45, a.1 ad m 6.
9. q.8, a.1.
10. q.68, a.1.
11. q.68, a.3.

CHAPTER 5

THE WEDDING OF HEAVEN AND EARTH

IN the often told story of Galileo, he has frequently been cast as the heroic antagonist of superstition and dogma, motivated by the desire for pure freedom from the trammels of the past; thus, father of the enlightenment, of free thought. But Galileo never claimed to be other than a good son of the Church, if sometimes a disobedient son. He never denied the authority of the pope: he simply wished him to use it wisely. The cause he championed was not free thought but free science. He wished to make the Church the champion rather than the antagonist of the Copernican Revolution. He saw no reason why the Church could not adopt the Copernican view of things, as it had once adopted the Aristotelian. After all, in the thirteenth century, Aristotle was viewed with the same hostility as Copernicus, yet the Church had learned to live with Aristotle. Aquinas, who had had to fight to have Aristotle accepted, was *the* theologian of the Church. Aquinas had shown the way to the adaptation of Scripture. Why should not the same method be used for Copernicus?

Galileo suggests the same stratagem as Aquinas in dealing with Scripture. In his letter to Grand Duchess Christina, he defends himself against those who would quote Scripture to affirm the Ptolemaic system and against Copernicus. Galileo has adopted the Copernican system wholeheartedly and become its protagonist and propagandist. But he has done so purely from scientific grounds and it is on these grounds that he wishes to have the matter decided by the Church, not on the basis of Scripture. For Scripture, says Galileo, is not a reliable guide in these matters. It is written in language which "avoids confusion in the minds of the common people''; it ''condescends to the common capacity''. This being so,

> I think that in discussions of physical problems, we ought not to begin from the authority of Scriptural passages, but from sense experience and necessary demonstrations; for the holy Bible and the phenomena of nature proceed alike from the divine Word. . . . It is necessary for the Bible, in order to be accommodated to the understanding of every man, to speak many things which appear to differ from the absolute truth so far as the bare meaning of the words is concerned. But Nature, on the other hand, is inexorable and immutable; she never transgresses the laws imposed upon her, or which necessary demonstrations prove to us ought not to be called in question (much less condemned) upon the testimony of Biblical passages which may have some different meaning beneath the words.

The idea that science, because of its exactitude, is more to be relied on than Scripture was not likely to endear Galileo to the Church authorities, even though Aquinas' interpretation of Genesis was not very different from just this. That Scripture could have several layers of meaning was an accepted idea in exegesis. But that science could show which was the exact and which the inexact meaning would hardly be a popular idea in the Vatican. After all, what was the teaching magisterium of the Church for if not for just such distinctions?

The cause for which Galileo was fighting shines through, however. He desired that scientific questions should be decided by scientific means and that the Church should abstain from exercising its teaching authority in this realm lest it teach as true that which science should show plainly to be false. The problem was that this was no longer a learned squabble among astronomers. The Copernican view involved not simply the matter of cpicycles. It upset an entire cosmology, a cosmology which had been woven into men's lives and into their faith. Men had learned to think of the earth on which they lived as a stable, unique and central platform. The Ptolemaic view lent itself well to the observation of the senses and the use of faith. Anyone with eyes could *see* the sun rise and set, could see the moon and the stars wheel their way overhead. It seemed obvious, furthermore, that there was a difference between things on earth and things in the sky. Here things changed: there they endured. A man could grow old but he never saw the stars grow old: they appeared to his dying eyes as they had appeared when first he saw them in his youth. They were symbols of the eternal, of a realm different from the earthly. It was a symbolism iterated daily in the liturgy when one prayed, "thy kingdom come on earth as it is in heaven". And when one read in the Scripture that "God had created heaven and earth", it is *this* earth surrounded by *this* heaven which was meant.

It was not primarily a matter of size, with the universe suddenly grown immense. Astronomers had always known that the distance to the planets from the earth was a very long way. The book which every educated man in Medieval Europe had read was Boethius' *Consolation of Philosophy*, and there Boethius says that "as you have learned from astronomy, the whole circumference of the earth is but as a point compared with the size of the heavens. That is, if you compare the earth with the circle of the universe, it must be reckoned as of no size at all." It was not a question of size, but of centrality and uniqueness, especially of uniqueness. For with Copernicus, the old distinction between "heaven" and "earth" disappeared. The earth was now in the heavens. The eternal and the changing had been merged. It was a point emphasized by Galileo when he showed with his telescope that the moon, far from being a perfect sphere, was pitted and scarred like the site of some ancient battlefield. The appearance of meteors and their location above the moon was also crucial. Such appearances showed that things do happen in the heavens, that they are not immutable. Sunspots on the sun indicated the same.

The change in thinking required in moving from a Ptolemaic view to a Copernican was thus considerable. Yet the surprising thing is that it was done with so little effort, all things considered. When Galileo was condemned, the battle was already over. The astronomers were all convinced. Indeed, they had been using the Copernican system for decades to make their calculations. From the time of Galileo's condemnation to the publication of Newton's *Principia* is a scant 50 years, and by that momentous event, debate had all but ceased on the issue of Copernicus versus Ptolemy and had shifted instead to Cartesianism versus Newtonianism, and both of these were systems which accepted the movement of the earth as axiomatic. As you read the authors of the time, you are surprised to find so little trauma over the new astronomy. Donne is widely quoted, and Pascal felt himself poised between infinities; but, apart from these two, you would look far and find little feeling of being lost on a small planet in the infinities of space. It is we moderns who have inherited such a *Weltangst* and not our seventeenth-century ancestors.

In short, apart from the condemnation of Galileo, the Church soon learned to make its peace with Copernican astronomy. The Church still prayed "thy kingdom come on earth as it is in heaven" and still stated its belief in God as "Maker of heaven and earth", but "heaven" had grown a little more vague in its meaning. There were new questions for theologians to grapple with, such as the possibility of men on other planets and the problem of their salvation. The location of heaven and God's throne became symbolic. Yet people still prayed: the creed was still recited. The Church learned that a cosmology may change and faith endure.

But there were other problems raised by the Copernican Revolution, subtle problems of motion and impact, which were to be much more serious in the long run, problems which engaged the best minds of the century. It was ultimately to be the problem of explanation. Were the new science and the categories it employed sufficient explanation of things? The categories of Aristotle could be used by Aquinas to point men to God. Could the categories of the new science be used in the same way? Would the new cosmology be as amenable to faith as the old when it was seen that it was not simply the matter of the earth's position but the way in which men thought of that earth? For, in the new science as formulated by Galileo, there was the germ of a whole new way of explanation, a way which was expanded by Descartes and culminated in the Newtonian system. It is this system of explanation which was to prove the great challenge to faith.

The new system of explanation took its roots in a new analysis of motion, an analysis begun by Galileo and made part of a metaphysical system by Descartes. For it is motion which is the Achilles' heel of Aristotle. A ball thrown, an arrow projected from a bow, what causes them to continue to move? The answer given by Aristotle was that, since all motion implied a cause continually exerted, somehow the air itself was set in motion with the throwing of the ball, and continued to act upon the ball, impelling it forward. For Aristotle, motion of any kind, either natural or "unnatural", requires a constant cause. In the case of natural motion, the object continues to seek its "natural" place of rest. In the case of unnatural motion, or a projectile, the air set in motion continues to push it forward.

Here, incidentally, is a reason why Aristotle rejects the void. In a void, there can be no air to continue the motion of a projectile. And, in a void, there would be nothing to impede the fall of a body moving naturally. Without such an impediment, a body would (said Aristotle) move at infinite speed, for its speed depends upon its mass and the resistance of the air. But such instant or infinite speed is an absurdity. Further, in a void, there can be no "structure" to space, nothing to make one place as compared to another. In a void, there is pure homogeneity, no articulation, no "structure". In a void, all directions would be equally "up" or "down" and a body would not know which way to fall.[1] A void would be pure geometry, the abstract space of Euclid and Archimedes, a concept no doubt useful in mathematics but not to be entertained in physics.

For Aristotle, motion is always a process, a becoming, a motion from the potential to the actual. A thing moves because it moves *toward* something, and that toward which it moves and away from which it moves is of the essence of that motion. In violent motion, a thing moves because it is being pushed by something else. Never does a thing move "on its own". The idea of motion as a state, requiring no other explanation than that same state, is foreign to Aristotle. Motion only endures so long as it is caused, and it is intrinsically different from rest. Rest is the end of motion. As the acorn seeks the sun and man seeks God, the attainment of a goal is the end of the seeking and the end of motion.

All of this is majestic, consistent, and wrong. Take a ball resting on a horizontal plane and give it a push. It rolls, and, if even and massy and on a smooth surface, it will continue to roll. What makes it move? Obviously not the wind, for the slightest touch will get it going. It is not in "natural" motion. Yet it is not in "unnatural" or violent motion either, for it takes no effort to keep it rolling. Then what state is it in? The answer must be that it is simply in that state of motion, and will continue in that motion until resistance brings it to a stop. Without resistance, it might well roll on, and on, and on. . . . If the plane extended far enough, it would never stop but would continue past the earth, moon, sun and stars to infinity, in its course uniting earth and heaven in one mathematical line.[2] Galileo found the moon pitted and spots on the sun. Heaven was no more perfect or unvarying than earth. The distinction between "natural" and "violent" motion and between earthly and heavenly physics was removed. The line of the body moving "simply because it moved" united all with the tool of mathematics. Before, this mathematical method had been reserved for astronomy, the perfect and geometrical realm of the stars. "Physics" was the "physics" of Aristotle, the study of moving, growing, changing things, and was thus appropriate for the earth as mathematics was for the heavens. Galileo wed the two in a mathematical physics which applied its analysis to falling bodies and, in a simple way, was able to relate the velocity of fall with the time of fall. It told nothing of the "reason" for the fall. It did not claim in its equation to give a cosmological and metaphysical explanation for the fall. It simply described the fall, correctly, and that was something which had not been done before.

The method of Galileo did not contain a metaphysics, but it could easily imply one. The rejection of final cause as a means of explanation in favor of efficient cause mathematically stated, the rejection of any "privileged" position in space, the new concept of time allied, not with change or becoming but with motion considered in the abstract, thus an independent parameter by means of which such motion could be measured: the stripping of substance of any secondary qualities not amenable to mathematical treatment — all of this was either implicit or implicit in Galileo, and took its place in the new conceptual scheme which came to replace that of Aristotle, namely that of Descartes.

Descartes is a mathematician: more importantly, he is a geometrician. He sees everything in terms of straight lines. With his earliest studies with the Jesuit fathers, he was convinced that mathematics, and mathematics alone, was the path to certainty.

> Those long chains of reasoning, simple and easy as they are, of which geometricians make use in order to arrive at the most difficult demonstrations, had caused me to imagine that all these things which fall under the cognizance of man might very likely be mutually related in the same fashion — for it has been the mathematicians alone who have been able to succeed in making any demonstrations, that is to say, producing reasons which are evident and certain.[3]

This certainty which we find in mathematics must, says Descartes, be the certainty we seek in the physical world as well.

> I do not accept or desire any other principle in physics than in geometry or abstract mathematics, because all the phenomena of nature may be explained by their means, and sure demonstration given of them.[4]

Geometry is the ideal means of investigation because it reduces all things to that which is clear and distinct. There are no fuzzy edges, no areas of indefiniteness, in geometry. What you see is what you get. "Matter" is no longer filled with scholastic "forms" but simply extension.

> The nature of body consists not in weight, nor in hardness, nor color and so on, but in extension alone. In this way we shall ascertain that the nature of matter or of body in its universal aspect, does not consist in its being hard, or heavy, or colored, or one that affects our senses in some other way, but solely in the fact that it is a substance extended in length, breadth and depth.[5]

Here anything "hidden" is stripped from matter, and, by reducing it to extension and geometry, all differentiation or heterogeneity in space is also eliminated. Now there surely is no "up" or "down". The privileged positions of the Aristotelian cosmos in which things fall because their "place" is at the center of the earth and fire rises because its "place" is in the heavens is here replaced by what Koyré called the "geometrisation" of space. Where formerly geometry had been an abstraction, a realm of thought for mathematicians touched by the real and everyday world only at selected points of navigation and surveying, it here becomes for Descartes the whole world, to be perceived by reason and intuition and able to explain in its own terms all things in earth and heaven, save God and the soul.

A. E. Burtt among many others has called attention to this separation in Descartes of the primary from the secondary qualities of matter, and claimed that this reduced the world to a bare and forbidding place, alien to the spirit of man.[6] In defence of Descartes, it should be noted that this also made of the world a place where one pattern of explanation could be used for many things. The scholastic philosophy could talk about fish and fowl, but there seemed little in common between them but a classification. The fish had a "form" that made it what it was, but an explanation of fish did not extend to that of fowl because that form was a different one. Descartes would deal with both as simply "matter", and when you explain the matter which compose them both, you are on the way to an explanation of both. For the apostle Paul, "there are celestial bodies and there are terrestrial bodies, and the glory of one is not the glory of the other".[7] But, for the new science, they were simply one thing — body. The same law could apply to both. Celestial and terrestrial, fish and fowl could be caught in one net, that of mathematical physics. Fish was body and fowl was body and both were extension and extension could be explained by the laws of geometry.

Living things were explained by Descartes and his followers in terms of bellows, valves, and pumps. Animals were automatons without feeling or sense, pulled and pushed as puppets. All the processes of life we explicated by mechanical means and efficient causality. The same explanation would suffice for them as for planets. Always you begin with matter in motion. The matter is broken into tiny particles, or "subtle matter" and into intermediate particles. The more solid particles form into clusters, about which sweeps the "subtle matter" in eddies, passing through it, pressing upon it. From this impact, interweaving and interchange, comes mountain and valley, star and planet, fish and fowl. Descartes' cosmology in its particulars is complicated, ingenious and very wrong. But its impact was remarkably clear: by stripping from matter any form and making of it pure extension, he can deal with everything in the same fashion: one size fits all.

All the occult and privileged is stripped from the world by the new definition of matter; all of the old penumbra of change, process and becoming, is stripped from it by a new definition of motion. Motion is no longer the change from something potential to something actual as with Aristotle and Aquinas with examples familiarly given from heat and cold, knowing and teaching, sickness and healing. It is now severely a change of place or position within a geometrical world. It is not a process but a state, a state of rest or a state of motion. Pick one such state; pick the other; it is a matter of indifference, for one is as likely, as "normal" as another. A body in motion continues in motion, indefinitely, in a straight line. Why should it stop? All places are alike to it, and it has nothing within to change it, containing neither memory nor anticipation, only bare extension. A body at rest remains at rest, forever, unless another body strikes it. Why should it move? One place is the same as another to it. It is inert, content.

The place of God in all this is equally simple and straightforward. God sets motion going and conserves it. That, and that alone, is his relation to the physical world. Contrariwise, the relation of the physical world to God is no longer one by which the mind may be led to God. For Aquinas, thanks to the potentiality/actuality analysis of motion, it is possible to go from the fact of motion to God as pure actuality. The order of explanation of motion is one which includes finality, for motion itself includes finality. Motion is change, becoming, *telos*. The explanation cannot proceed to infinity because of what explanation means, i.e., to explain motion is to set forth its end, what it is all for, where it is going.

It is not so for Descartes. There can be for him no hierarchy leading upward to God from motion and physical cause, for he has stripped from them anything by which to begin such a climb.

> I have not drawn my arguments from observing an order or succession of efficient causes in the realm of sensible things, partly because I deemed the existence of God to be much more evident than that of any sensible things, partly because this succession of causes seemed to conduct merely to an acknowledgement of the imperfection of my intellect, because

> I could not understand how an infinity of such causes could have succeeded one another from all eternity in such a way that none of them has been absolutely first. For certainly, because I could not understand that, it does not follow that there *must* be a first cause, just as it does not follow that, because I cannot understand an infinity of divisions in a finite quantity, an ultimate atom can be arrived at, beyond which no further division is possible.[8]

The order of efficient material causes leads Descartes, not upward to God, but out into the infinity which stretches before or after in time, or into the infinity of extension and division. It is an infinity which he cannot grasp, nor be sure that it can ever have a limit. He will not, to be sure, say that extension is infinite, but certainly it is "indefinite" in the sense of having no limit. The fact that his understanding is limited by no means implies a limit to either extension or division.

The way to God begins, not with material things, but with the soul.

> Therefore I prefer [we continue the above quotation] to use as the foundation of my proof, my own existence, which is not dependent on any series of causes, and is so plain to my intelligence that nothing can be plainer: and about myself I do not so much ask, what was the original cause that produced me, as what it is that at present preserves me, the object of this being to disentangle myself from all questions of the succession of causes (causes "in the realm of sensible things"). Further, I have not asked what is the cause of my existence insofar as I consist of mind and body, but have limited myself definitely to my position insofar as I am merely a thing that thinks.[9]

For Descartes, a proof of God cannot be based on the things of sense, on the everyday and common fact of change as it is in Aquinas, for with the new science, the things of sense have been stripped to extension and motion. There is in these things no trace remaining of the divine presence by which the mind may be drawn to God. They are simple, reduced, bare, serving admirably for the rask of mathematical explanation but of no value for intimations of God. For that, one must turn inward, for God is to be found not in the world but in the soul.

It is from this certain knowledge of God, discovered in the soul, that the laws of physics are themselves known. For Descartes, one approaches the world of nature not from things of sense, as with Aquinas, but from the mind to God to determine that which is sure, and then from God to things of sense, to nature. One can with assurance reduce the world of nature to extension and motion and reason from clear and distinct ideas because God is the guarantor of those ideas. One comes to nature with ideas verified and ready to be applied, stamped with God's approval. And, from that nature of God, a God infinite and unchanging, it is possible to deduce yet other truths concerning the world of nature. These are truths concerning the law of motion, which is the simple law that God "always preserves the same amount of movement in the universe".[10]

> God, who created matter together with motion and rest by his ordinary action, always preserves that same amount of motion and rest. . . . We are thus able to see the perfection of God in that not only is he himself immutable but always acts in the same constant and immutable manner.[11]

By "quantity of motion", Descartes makes clear that he means what we would express as momentum, or mass multiplied by velocity.

> If one portion of matter moves twice as fast as the other, but the other is twice as big, there is the same quantity of motion in each.[12]

Here the law of inertia, so effectively used by Newton, is given by Descartes its first modern formulation, derived from the immutability of God and matter as pure extension. Descartes puts it: "nothing happens without an external cause". Since God immutably keeps the same quantity of matter in nature, and since there are no hidden forms which cause things to change (for always for Descartes what you see is what you get), the only change which can occur in nature must be that brought about by an external cause.

> If a part of matter is four-sided, it is easy to see that that matter will remain four-sided, unless something comes to change it. If it is at rest, it will not move unless something causes it to move.[13]

If there is no internal cause of motion, then only an external cause can produce a change. A square piece of matter remains square until something changes it from that shape. There is for Descartes no explanation for being four-sided other than that in the interchange of parts and motion of things, it has become so, and will now remain so unless its parts are changed again. That which is in motion will remain in motion; that which is at rest will remain at rest; and God who is immutable will immutably keep it so.

It is by a similar train of thought, plus an atomistic view of time, that Descartes will conclude that a body in inertial motion will move in a straight line. For Descartes, there is no "carry over" from one moment to the next. Each moment is individual. This atomistic view of time is given in a famous passage of the *Meditations*. Descartes has been arguing for the existence of God from the idea of God.

> But though I assume that perhaps I have always existed just as I am at present, neither can I escape the force of this reasoning, and imagine that the conclusion to be drawn from this is, that I need not seek for any author of my existence. For all the course of my life may be divided into an infinite number of parts, none of which is in any way dependent on the other; and thus from the fact that I was in existence a

> short time ago, it does not follow that I must be in existence now, unless some cause at this instant, so to speak, produces me anew, that is to say, conserves me. It is as a matter of fact perfectly clear and evident to all those who consider with attention the nature of time, that, in order to be conserved in each moment in which it endures, a substance has need of the same power and action as would be necessary to produce and create it anew, supposing it did not yet exist. . . .[14]

Time is here no longer a part of process and change; it has no future as it has no past. Each moment implies nothing but itself and God and a next moment is derived only from God.

So, in the laws of motion, each body tends to move, not in a circle, but in a straight line.

> Parts of matter, as we observe, do not tend to move in any line at an angle, but always straight ahead. . . . The cause of this is the same as before, namely the immutability and simplicity of operation, by which God conserves motion in matter. He will conserve in it nothing but the precise quantity which it had in that moment, not as it might have been previously . . . thus always in a straight line and never in a curve.[15]

The body moves in a straight line because of the immutability of God, and the motion is conserved, "not as it might have been some time previously, but as it is at the precise moment he conserves it". That is, the motion has no memory. As it is at the precise moment, the instant, that is how the motion will be conserved. Descartes gives the illustration of a body moving along a curved path, as a stone in a sling. At the moment it is released, it will continue along a straight line tangent to the curve at that point. Only by constraint can it be kept in a curved line. The motion conserved apart from that constraint will be the motion at the precise instant it is released. And this happens because of the way in which God chooses to conserve motion, namely, from instant to instant. The past history of the particle has no influence upon its future history. Only its immediate condition counts.

It is thus from the immutability of God and a moment-to-moment creation that Descartes describes the law of inertial movement in a straight line, just as from the immutability of God and the nature of extension he derived the law of inertia itself.

The same reasoning is used for extension as is used for the mind. Examining myself I find myself simply a thinking being as a stone is an extended being. There is no other form in the stone than extension and no other attribute in me than thought. In neither the extension of the stone nor in the thought is there any "power" or "force", no hidden or mysterious depth which would cause me to be in the next moment as I am now. Thus, only by the continued creation of God is the motion of the stone in one case and my own existence in the other conserved from one moment to the next. The stone is re-created with its momentum unchanged and I am re-created with my memories intact. Since neither extension nor thought can contain within their crystalline structure anything unknown, the moment too can contain only itself, be the repository of no memories, the anticipation of no future.

To this view of extension, Leibniz will take exception, but his criticism will begin with the laws of motion as Descartes derived them. So it will be necessary to take a closer look at those laws. Since God, according to Descartes, conserves the same amount of motion in the world, and since motion leads to motion only through impact, that quantity of motion must be preserved in impact. This, for us, is simply mv, mass multiplied by velocity, except that Descartes does not have our modern (Newtonian) concept of mass. It is, for him, simply extension. Further, for Descartes, motion is itself an absolute, not in the Newtonian sense of being measured against absolute space, but in the sense that it is a quantity in and for itself alone. It is to be known and measured simply by relation to other bodies. An intervening and growing gap between A and B can be called either A moving away from B or B moving away from A. Motion is not in itself motion "toward" anything. Direction is irrelevant to motion. It is not a directed quantity, as it will be for Newton. It is a scaler, not a vector. Motion for Descartes is not and never can be allied with any sense of process, change, or end. It is measured as an "absolute" without respect to its direction.

Thus, to compare one "quantity of motion" with another, one simply takes the mass (or extension, or weight — Descartes is ambiguous as to how this is to be measured, whether with a scale or a ruler) and multiplies it by the velocity. It is a simple geometrical rectangle with motion as one line and extension the other, and the quantity of motion measured by the area of the rectangle. When two bodies meet, assuming that they are each perfectly hard, one can, by this quantity of motion, decide what their effect will be upon each other. Quite simply, the one with the larger quantity of motion will prevail. If A is larger than B and A meets B, then whatever it is that A's direction is, it will prevail and carry B with it so that the "quantity of motion" of the two is conserved. If A meets B and B is at rest and larger than A, then B's state of rest will prevail, since inertia is a form of motion or persistence — and A will be reflected from B. To be in motion or to be without motion requires no outside reference, is an absolute which continues to be well overcome by some other absolute equally persistent in *its* state of being. If A in motion meets B in motion, each with the same extension and the same velocity, then, since neither can prevail, each will rebound and each will keep that same quantity of motion.

All this is simple, geometrical, rational, clear and wrong. Only the rule about equal masses approaching with equal velocities turns out to be true: the others are false. But one can at least see why they are false. It is, again, because of Descartes' geometrical approach. All must be clear and precise, black or white. A body is smaller than, equal to, or larger than another, and according to that difference, and no matter what the extent of the difference, its resulting motion will be determined. There is no place for shadows here, no nice shading off from one case to another. Descartes' is a straight line philosophy.

The trouble is that straight lines do not occur in nature. No body ever moves in a perfectly straight line unless it is the only body in the universe. Such linear motion is a mathematical fiction. Descartes in part realises that his laws of impact are equally fictitious. "They will occur", he says, "only where no other bodies touch the two that are involved, and where the bodies themselves are perfectly hard." But since Descartes' universe is filled with a *plenum* of "subtler matter" which surrounds all and interpenetrates all, such abstract conditions will never exist. Descartes has thus proposed rules of motion which his own cosmology excludes as possibilities.

The external relation of God to the world is clear in Descartes' view of the so-called "eternal ideas". According to the traditional teaching of the schools, the eternal ideas had in them something of God, still had about them the aura of the neo-Platonic realm. For neo-Platonists and most of the medievals, the eternal ideas proceeded from God as "rays from the sun". Thus the eternal ideas were another means by which the world was united to God, for in thinking eternal truths, said the medievals, we share in the thoughts of God. The eternal truths, being part of the nature of God, could not be other than they in fact are.

For Descartes, however, the eternal truths are created by God and could be other than they are. God could, for example, have created a world in which triangles would not have three angles and mountains could exist without valleys. The reason why Descartes maintained this must be because he wanted to found a physics upon these eternal ideas or essences. But to do this, they must be able to be grasped by the intellect, must be clear and distinct and available for the mind.

> For the knowledge upon which a certain and incontrovertible judgment can be formed, should not alone be clear but also distinct. I term that clear which is present and apparent to an attentive mind . . . but

> the distinct is that which is so precise and different from all other objects that it contains within itself nothing but what is clear.[16]

Again we meet the requirement that there be nothing hidden, nothing needing completion in anything but itself. There is for Descartes no hierarchy of truth with all truth partial that is not summed up in God. Rather, everything is clear and distinct, for it is upon such ideas that knowledge is to be built.

It was not possible for Descartes to set up these truths independently of God as though they would exist even if God did not. Rather, these truths are dependent upon God, part of the created world, not part of God. What they lose in absolute certainty they gain in availability. As substantial form is stripped from things to make them amenable to mathematical physics, so too eternal truth is shifted from a realm which rises to God to one confined to this world. They are longer signs of God but of the new science.

Since eternal truths are created as contingent things, they can carry their own conviction without being part of an uncreated realm. Because they are contingent, they are discreet and separable. A science can be erected upon them which is certain because God has willed the truths to be as they are and God can be counted on. But again, it is but a step to say that, since we can count on the truths, we can do so apart from reference to God, even as we can know them without such reference. A world, thus, other than God and from which our reference to God can be stripped can become a self-explaining world, and the science given its charter by Galileo here is on the road to leading its own life apart from God and explaining itself apart from reference to God.

This also means that the world is arbitrary in the deepest sense in that God could have made everything, even the truths which seem irrefutable, to be other than they are. The laws of logic, the laws of mathematics, could have been different, and to the question "why did God make them as he did?" there is no answer but simply "because he willed thus to do".[17] There is a fully rational world able to be penetrated by the mind: science is given its charter but to the question "why are things so and not otherwise?" there is no answer to be given.

We have claimed that it is in this new understanding of the physical world, in particular, the new understanding of motion, that the real challenge lay to Christian faith. But how can this be, since Descartes speaks so much of God, indeed, derives the very laws of physics from him? The problem is simply this: that, for Descartes, God is external to the world, and rules it only by push and shove. He is nowhere in its depths, urging it to any completion or providing for it any future. The push and shove can be codified in laws of motion, and, once this is done, the world can become a self-explaining whole, functioning "on its own". Once we know that motion in a straight line is the fact, do we really need God to explain that fact, since the state of motion is sufficient to itself? For Descartes, it is true, you can get from one moment to the next only by invoking God. But, since God in fact conserves the motion of the one moment into the next, it is a short step to take that fact simply as a "given" and leave God out.

The world thus can become a self-perpetuating and self-explaining whole, needing God, if at all, only to get it going, and, from the world to God, there is no way except to the divine cue stick which first begins to rattle the billiard balls. That first motion given, all else follows.

Thus with Descartes there becomes possible a new thing: a cosmology without God. The wedding of heaven and earth in the new schema of mathematical physics leaves no physical place for God's presence, and no way for him to work in the world save by intervention. The line of mathematical motion is a straight one that binds together heaven and earth. Such a line has no special or privileged points. One place is like another. And the line points on, and on, to infinity, but not to the infinite. No matter how far it goes, it is simply place extended, thus points not to God but only to itself. The clear light of reason bathes all in its light, leaving no dark corners, no hidden depths in which God can work in silence and in mystery.

The challenge is one of explanation. Was the new science a sufficient explanation of things? Was the mathematical–physical wedding of heaven and earth sufficiently wide-reaching and subtle to make the concept of God obsolete in all scientific inquiry? But this was a subtle challenge, one proposed in the categories of space and time, motion and force, one barely understood even to those who engaged in its polemics. The one who did seem to understand was Gottfried Wilhelm Leibniz. The trouble is that almost no-one understood *him*. Perhaps, looking back at the struggle with the advantage of years of perspective, we can be more perceptive than his contemporaries, can see in contrast to the straight line world of Descartes the one that Leibniz will propose, one full of intricate weavings and bendings, with alternatives at each turn and an unborn universe hidden in the shadows.

NOTES

1. Alexandre Koyré, *Études Galiléennes*, p. 23.
2. *De Motu*, quoted in Koyré (*op. cit.*), p. 77.
3. Haldane and Ross, *Descartes: Philosophical Works* (Cambridge, 1911), I, p.93.
4. *Op. cit.*, I, p. 269.
5. H. R., I, p. 256.
6. A. E. Burtt, *The Metaphysical Foundations of Modern Physical Science* (1924).
7. 1 Corinthians 15:40.
8. H. R., II, 13.
9. *Ibid.*
10. *Principles of Philosophy* #36.
11. Adam and Tannery, Paris, 1897, VII, 61.
12. *Ibid.*
13. A. T., VII, 62.
14. A. T., IX, 39.
15. A. T., VIII, 64.
16. H. R., I, 237.
17. H. R., II, 248.

CHAPTER 6

THE MATHEMATICAL WORLD OF GOTTFRIED WILHELM LEIBNIZ

LEIBNIZ once said that his whole metaphysical system could, "in a manner of speaking", be expressed in terms of mathematics. It is in his discovery of the infinitesimal calculus and his discovery of "force" in dynamics that one can trace the basis of that metaphysical thought. It is important to do so, because it is only on the basis of his understanding of the world that we can grasp Leibniz's own version of the cosmological argument and his perception of the relation of the world to God in creation.

In his essay on Descartes' principles, Leibniz puts forward a concept which, he claims, goes beyond Descartes and shows Descartes' laws of motion to be in error and, indeed, superficial. It is the concept of "force" or *vis viva* or what we today would call "energy". Descartes measures "quantity of motion", or what we would call momentum, as the basic quantity in motion, where motion is given in absolute terms. Thus, consider one body moving at four feet per second and weighing one pound and another moving at one foot per second and weighing four pounds. According to Descartes, their "quantity of motion" would be the same, thus they would be equivalent in this basic category. But this is not to look at them closely enough, says Leibniz. There is something about them which is much more basic. Consider them, not at the point of impact and the fraction of time in which they collide, but consider them in the result of their motion. Let each fall from a different height. Let the one pound weight fall from four feet and the four pound weight from one foot. Now what do you have? Equivalency? Not so. For, according to Galileo, the velocity of a body is proportional, not to the height from which it falls, but from the square root of that height. Thus, the four pound weight will have a velocity half as great as that of the one pound weight, so that its "quantity of motion" will be twice that of the one pound weight, i.e., its m will be four, but its v, only one, thus its mv, only four, while the one pound ball, having reached a velocity of two will have an mv of two. In this case, something more than Descartes' "quantity of motion" is involved. What is equivalent is their "energy", or m times v squared in each case.

Now this "force" is not something immediately observable, says Leibniz. You see its results (and Leibniz illustrates with two balls suspended from cords striking one another) but could not (claims Leibniz) calculate it from clear and open observation, as Descartes claims you can do for quantity of motion. There is something here hidden from the eye, a mathematical quantity but not a geometrical one.

Of course, both Descartes and Leibniz were right. Momentum is conserved if you figure in the direction of motion, as is energy. But the point that Leibniz would make here is that energy or *vis viva* is not to be calculated by geometry and yet seems a more basic quantity than Descartes' "quantity of motion". It seems that in this case you get more than you see. Further, Leibniz feels that this "force" or energy is able to be related to a whole new understanding of motion that goes far beyond Descartes. But, to understand this, we must go back to Leibniz's great mathematical discovery, the infinitesimal calculus.

In 1694, Leibniz published an essay in which he put forward a method to express, not only a curve by an equation in the fashion of Descartes (inventor of Cartesian co-ordinates), but every point on a whole series of curves. He did this by giving a general equation of a curve in what we would call parametric form. In this equation some of the coefficients would be constant, others variable. By varying the variable coefficients, a whole family of curves could be produced, of which the equation itself would be the "law of the series".

> But comparative curves of the series can pass over one to another, with some coefficients constant or permanent (which remain not only in one, but in all series of curves), others are variable. Thus to give the law of the series of curves, it is necessary to have one of the coefficients variable to some extent. . . .[1]

As an illustration, Leibniz gives the equation $(x-b)^2+y^2=ab$ as such a general equation, in this case of a circle, with variable parameters a and b. He sets himself the problem of finding the common tangent of this series of circles, which he discovers to be a parabola. He says of this general equation that it is the "primary equation of all of our circles and of each common point"[2]. In this case, one of the variable coefficients will give the size of the radius of the circle, the other, the position of the circle. By varying both coefficients, one is able to produce an entire series of curves of varied size and location, of which each individual curve can be determined and each individual point on each curve. Thus, an infinite number of circles can be produced from the one general equation or "law of the series" of circles, each circle varying by an infinitesimal amount from the next by the variation of one of the variables. The tangent common to each of the circles, or the "envelope" of the family of circles, can be obtained by differentiating with respect to the variable parameter.

Now it is in this "law of the series" that Leibniz finds the secret of his "force". Each body in motion is describing some curve which can be expressed in an equation. It is the equation, not the motion, which is primary. "Force" is the revelation of this law of the series, and motion is the result of the force. Motion, which for Descartes is the primary and absolute quantity, is derived and secondary to force. Leibniz also introduces a "derivative" force which he distinguishes from "primitive" force. Both, however, are to be distinguished from motion. In a letter to DeVolder he says,

> You assert that motion, or the product of mass and velocity, constitutes the derivative forces. I, however,

> do not consider motion to be a derivative force but think rather that motion, being change, follows from such force. Derivative force is itself the present state when it tends toward or pre-involves a following state, as every present is great with the future. But that which persists, insofar as it involves all cases, contains primitive force, so that primitive force is the law of the series, as it were, while derivative force is the determinate value which distinguishes some term in the series.[3]

Here Leibniz relates force, his discovery in dynamics, to the law of the series, his discovery in calculus. "Primitive force" is evidently the basic equation which includes all the possible curves; "derivative force" is derived from this equation, perhaps an equation of a particular circle. Motion is the result of this derivative force. At any given point, it would be a straight line, tangent to the curve as found by taking the derivative at that point. It is this last, the straight line tangent, upon which Descartes has fixed his attention, when it is the much more important "law of the series" which gives the true understanding.

Notice how those things which the new science had stripped from things are now returned by Leibniz, dressed in the acceptable garments of mathematics. For Descartes, substance is purely inert, has no substantial form for its essence, its extension, has no goal for all place is alike to it, has nothing hidden within it yet to become. Motion, for Descartes, has no memory and anticipates no future. Process or change is simply a shuffling of parts. Motion or rest are equally indifferent to such a substance. A ball released from a string swung around the head will travel in a tangent to the circle for it has no remembrance of the past. Each instant is separable from each other instant.

All of this follows from reducing material substance to extension and motion to change of place. But, for Leibniz, material substance is not extension but the law of the series. Motion is not change of place but the visible result of force, itself derived from the law of the series. This law is to material substance what substantial form was to the matter of Aristotle and Aquinas. Leibniz calls it an "entelechie". It is an "internal tendency to change" which he claims is in all substance.

> The case is like that of mathematical laws of series, or the nature of curves, where the entire progression is sufficiently contained in the beginning. Nature as a whole must be like this: otherwise it would be absurd and unworthy of wisdom.[4]

> There is a tendency, or a spontaneous progress, within all substance. It is that force or tendency which I can call by no better name than that of Entelechie.[5]

It is in this "force", able to be mathematically expressed, yet related to the very metaphysical substance of things, that Leibniz finds his clue to a metaphysics different from that of Descartes.

> This consideration, in which force is distinguished from quantity of motion, is of importance not only in physics and mechanics — in finding the true laws of nature and the rules of motion — but also in metaphysics for the better understanding of principles. For considering only what it means narrowly and formally, that is, a change of place, motion is not something entirely real . . . Now this force is something different from size, figure and motion, and from this we can conclude that not everything which is conceived in a body consists solely in extension and its modifications as our moderns have persuaded themselves. Thus we are compelled to restore also certain beings to forms which they have banished.[6]

From a view of matter as extension and motion, only as change of place, Descartes has no way to rise to God. Is it otherwise with Leibniz? Can he relate his deeper concept of motion to God? The answer to the question is in his essay on God entitled "The Radical Origination of the Universe".

In this essay, Leibniz raises the question, to become *the* classic question for the cosmological argument, "why is there something and not nothing", or "what is the reason for the existence of the world"? This is usually assumed to be a general question which can be raised for any and all cosmologies. But it should be seen that Leibniz raises the question with the background of his own cosmology, a cosmology which has as its center his new concept of motion. To do so, look first towards the end of the essay where Leibniz clearly states this.

> We observe, indeed, that everything that happens in the world follows the laws of the eternal truths, which belong not only to geometry, but equally to metaphysics. This means that whatever happens does so not only according to material necessities, but also according to formal reasons. This is true, not only in general, when it is a question of explaining — why a world exists rather than no world, and why it is as it is and not different — a reason which certainly must be drawn from the tendency of all possibles toward actuality; it is equally true when we now step down to special cases. We shall then understand that everywhere in nature the metaphysical laws of cause, potency, and action apply in an admirable way and that they even prevail over the very laws of pure geometry which determine material processes. *This I understood with great admiration when I tried to account for the laws of movement.*[7]

Now read for "geometry" the philosophy of Descartes and for "metaphysics" that of Leibniz himself. With Descartes, we have an explanation in terms of "material necessity" only; but with Leibniz, one in terms of "formal reason" as well. Both Descartes and Leibniz use "eternal truth", the difference being that for Descartes, "eternal truth" is something God could change if he wished, making mountains without valleys and triangles without three sides, while for Leibniz, "eternal truth" is of the nature of God himself and thus not arbitrary. Now look again at Descartes' answer to the question "why does a thing move?" It is to be found, according to Descartes, in some other bit of matter, itself in motion, and from this bit of matter is another, and so on to infinity. From such an explanation there is no way to God except as the one who gives the first bit of matter *its* motion. The explanation is purely by "geometry", i.e., matter as extension alone. It is thus a "physical explanation" that acts by "necessity", i.e., without any goal or telos. For Leibniz, by contrast, the answer to the question "why does a thing move" is given in terms of force, which is itself given in terms of the "law of the series" and which is in turn given, in this essay, a final metaphysical explanation in terms of creation.

To the question "why do things move as they do?" Descartes will answer "because God made them to move that way", that is, by geometrical laws of motion which are themselves arbitrary, for God could have made the eternal truths other than they are. To that same question Leibniz will reply, "because they are part of this best of possible worlds, a world framed not by arbitrary decree, but according to the eternal truths which are not arbitrary, but part of the nature of God". Or, the explanation of motion is in the law of the series by which it moves, and the explanation for that series is to be found in creation.

Leibniz begins his essay with the illustration of a book, one copy of which has been copied from another. No matter how far back you go in the series of reproduction, you "can never arrive at a complete explanation, since you will always have to ask why at all times these books have existed, that is, why there have been any books at all, and why this book in particular?" In short, "neither in any single thing, nor in the total aggregate and series of things, can the sufficient reason for their existence be discovered". This applied to the states of the world as well. No single state, as no single copy of the book, can be a sufficient explanation of the whole series of states. There must, therefore, be some being apart from this series of states who is the "ultimate reason of things". This must be true even if the world be eternal.

It seems at first glance that we are here back with Descartes and one bit of matter pushing another to infinity. But this is not really the case. It is obvious that Leibniz is referring in his phrase "series of states" to the series which he found so profitable in his dynamics. For he continues,

> Things eternal may have no cause of existence, yet a reason for their existence must be conceived. Such a reason is, for immutable things, their very necessity or essence: while in the series of changing things, even though this series itself may be supposed to be eternal, this reason would consist in the very prevailing of inclinations, as will become clear soon. For in this case reasons do not necessitate (that is, operate with absolute or metaphysical necessity, so that the contrary would imply contradiction), but only incline.[8]

Here the contrast is between the series of states of things and eternal things. Even for things that are eternal there is a "reason", though they exist of necessity. Doubtless Leibniz refers to eternal truths, such as the angles of a triangle being equal to a straight angle, a favourite illustration of Spinoza. In the "series of changing things", here the actual state of the world, these reasons do not "necessitate", as do the eternal truths, but only "incline". In the next paragraph he refers to the "chain or series of things, the aggregate of which constitutes the world" and says that in this series is a "physical or hypothetical necessity which determines the later states of the world by the earlier" and that we must pass from this to "something endowed with absolute or metaphysical necessity".

Here the "series of states" is distinguished from "absolute necessity". It has instead "hypothetical or physical" necessity, thus does not contain its own sufficient reason for existence. In other words, the "series of states'' is contingent. It might not be, might be other than in fact it is. But it is not contingent in the sense of Descartes, i.e., it is not simply arbitrary. It may not ''necessitate'' but it ''inclines''. In short, this ''series of states'' of the world seems to be the same as the ''law of the series'' by which Leibniz would explain motion. It is neither necessary in the sense that it could not be otherwise, nor is it simply arbitrary, for a reason can be given for it.

In short, the ''series of states'' of the world seems to be a loaded term for Leibniz, loaded in the same direction as the term ''motion''. For Leibniz, motion is the result of force which is itself derived from the law of the series. It is not exemplified in impact and the picture of billiard balls hitting one another endlessly, but in a curve having its shape written within it and moving by that inward agenda. From this concept of motion, Leibniz can move to metaphysics, to creation, to God as Descartes could not do.

Notice how Leibnitz explains the act of creation which brings about this ''series of states'' of the world.

> Let us explain somewhat more distinctly how, from the eternal or essential — that is, metaphysical — truths, the temporal or contingent — that is, physical — truths are derived.

The physical and contingent world is to be derived from eternal truth. The world is to be contingent, yet not arbitrary.

> From the fact that something rather than nothing exists, it follows that in possible things, or in their possibility or essence itself, there is a certain demand or (so to speak) a claim for existence; in short, that essence tends by itself toward existence.[9]

There must be a reason why there is something rather than nothing, for God does not act arbitrarily.

A side look at the *Theodicy* may help here, where Leibniz

says that if God had decreed to make a material sphere, but with no reason for making it any particular size,

> this decree would be useless. . . . It would be quite another matter if God decreed to draw from a given point one straight line to another straight line, without any determination of the angle, either in the decree or in its circumstances. For in this case the determination would spring from the nature of the thing, the line would be perpendicular, and the angle would be a right angle, since that is all that is determined and distinguishable.[10]

In another place, Leibniz says that if "there were no best world among all possible worlds, God would not have produced any", for there would be no reason for God to pick one rather than another, and God does not act without reason. That which God makes must be complete, distinct, fully formed, fully individuated. There must be nothing amorphous about it, nothing optional, nothing which might be either added or left out.

Another look at our "law of the series" may clarify this. A mathematical model will help. The world for Leibniz is made up of substances which are describable by individual equations which give, in their variable parameters, a complete law of all events in the motion of that substance. Leibniz calls such a substance a "monad". This is the "entelechie" of a "complete substance". Each such monad is independent of all others, but coordinated with it in a pre-established harmony. Our model is thus made up of a set of equations, each equation with variable parameters. The total set of these equations is the "series of states of the world".

Now of each equation, it can be said that it does not exist of such necessity that its denial or variation is a contradiction. It might very well be altered, slightly or radically, but it cannot be so altered and remain in this particular set of equations, for the equations are bundled together in their pre-established harmony. Another way to say this is to say that the equations are not independent. You cannot alter one without altering all the others. Yet this particular set is not the only possible set. Indeed, there is an infinite number of other possible sets, differing from the given set by different degrees. Each of these possible sets is a possible world. There is, doubtless, a possible world for which it is true that Caesar will not cross the Rubicon, but it is a world with other variations to accommodate this change. A slight modification in any one of the equations will produce a new set of equations also modified. Each such modification will be a possible world.

Each such modification is a completely specified possible world, containing fully specified individual substances. So, in a discussion with Arnauld over many possible Adams, Arnauld complains that you cannot have several Adams who are still Adam. Leibniz agrees, saying that he is speaking of Adam from a human standpoint, thus of a general Adam, one not completely specified. They are both speaking of

> some person conceived *sub ratione generalitatis*, under circumstances which seem to us to determine Adam to be an individual, but which do not in truth do so sufficiently; as, for instance, when we mean by Adam the first man, whom God puts in a pleasant garden, etc. . . . But all this does not sufficiently determine him, so there might be several other disjunctively possible Adams. . . . But that concept which determines a certain Adam must include, absolutely, all his predicates, and it is this complete concept which determines the *rationem generalitatis ad individuum*.[11]

Thus, God, in creating this world, chooses it from among all those possible worlds which include completely specified Adams, so that Adam is not the same in any two possible worlds. Only by "abstracting" from this is it possible to speak of several possible Adams and be thinking of the same individual in each case.

In the same letter, Leibniz, supporting his statement that his future journey is already contained in his own concept, says, "If I did not make the journey, there would therefore be a falsehood which would destroy my individual or complete concept, or what God conceives of me or did conceive of me before he resolved to create me". Even before his creation of Leibniz, God sees Leibniz's future journey as part of that completely specified world which as yet remains only possible. And God sees as well other possible worlds in which someone *we* would call Leibniz, but is actually another person, does not make this journey. But God chooses, not those other possible worlds, but this actual world.

Thus, for God in creation, real alternatives are involved. If, in the possible world, there is something left vague or uncertain, then that is not the world God can choose. Only that which is completely specified, "concrete", can come into existence.

To return to our model, each set of equations provides a complete and specific alternative or possible world. That which is true of all sets of equations, or all possible worlds, is "eternal truth". Following our model, this would be the properties of numbers, geometry, etc., all of which must remain invariant for all sets of equations. These "eternal truths" are distinguished from any given set of equations. The particular set of equations plus the eternal truths make a possible world.[12]

The difference between these eternal truths and the possible worlds is the difference between the abstract and the concrete, between number and that which is numbered, between space and time taken as an order of possibility and space and time as a relation between actual existents. Into the abstract order any number of possibilities may be placed. Thus the abstract cannot of itself determine a complete individual, but remains simply the condition of complete generality.

> Space and time together are the order of possibilities of the entire universe, such that they order not only

> that which actually is, but also that which could be put in its place, just as numbers are indifferent to that which is numbered.[13]

This mathematical model of Leibniz helps us to understand two difficult points in his metaphysics: the ideality of space and time and the identity of indiscernibles. Leibniz has a long discussion with Clarke, a friend of Sir Isaac Newton, over the role of space and time. Newton upheld an absolute view of space and time, claiming that they existed independently of any and all events and were thus primary, as they are for Descartes. Leibniz wants to maintain that both space and time are secondary, derived from the monads, or law of the series. The trouble is that Leibniz makes space and time part of necessary truth, which seems to make them primary indeed. How can space and time not be absolute if they are part of necessary truth for Leibniz? The answer, according to our model, is that they are part of the eternal truths like those truths which are invariant in all possible worlds, and thus are not able *in themselves* to individuate and distinguish either one world from another or one complete substance from another. Space and time are for Leibniz in his discussion with Clarke just such ideal things, conditions, as it were, under which the possible worlds come to be.

This is also the point about the identity of indiscernibles. Space and time are not sufficient to individuate, to mark off one individual from another. Only the equation itself, law of the series, is adequate for that, for the complete substance *is* the law of the series. That which obtains for all equations and all sets of equations is the abstract, the eternal truths, within whose conditions the equations can indeed be stated, but which are of themselves inadequate for complete individuation.

Only in geometrical and "abstract" matters can the abstract and eternal suffice to individuate, and thus only in matters purely geometrical can two identical objects exist.[14]

There is similarity here to the question of several "Adams". We can think of several similar Adams only because we abstract from the fully concrete. In the fully concrete, two individuals are never identical: that can only happen in the abstract.

The difference is between the geometry of Descartes, an abstraction which never occurs in nature, and the fully concrete in which that abstract can be embodied. Only the law of the series, the equation itself, is fully specified, and two such equations cannot be identical for then they would be the *same* equation. The same is true of the set of equations that makes up this or any other world; no other such set could obtain, for it would simply be the *same* set. So Leibniz, answering Clarke that the world might have been created at an earlier time, replies that such a beginning "whenever it was, is always the same thing".[15] That is, unless it enters into that set of equations which goes to make up this world, nothing is altered. The same is true for any alteration in size. If everything were to be reduced in size, the world would be the same so long as the set of equations that goes to make it up remained the same. Shifting things in space and time makes no difference.[16]

All of this background is needed as we seek to understand Leibniz's concept of creation, for it is a creation of the world as Leibniz saw it with which we are dealing. Always, when we speak of God creating the world, it is of *this* world, the world as we see it and know it, that we speak. It is no different for Leibniz.

There is no doubt but that Leibniz accepted the new science. He criticises Descartes as one who follows Descartes and accepts the new view of things that comes from Copernicus and Galileo. He has no intention of returning to the old two-storey view of things. Mathematics is, as with Descartes, to explain all. But it is a different mathematics, a more subtle one, one that is to that of Descartes as the concrete to the abstract, the dense to the empty, analysis to geometry, the infinite to the finite. For Descartes, all is ruled by vision, by the clear and distinct.

Motion and extension, being just such distinct and geometrical qualities, are primary and all else is to be explained by them.

For Leibniz, however, while space and time remain mathematical and geometrical qualities, they are secondary, reflecting the outer and straight line motion of bodies as they are pushed by other bodies. More basic is the inner equation, the programmed history written into the heart of things, the law of the series.

The relation of God to this cosmos is not simply that of one external to it. He has, as it were, written himself within it. For Descartes, the laws of motion are external and arbitrary. For Leibniz, they are internal and framed by wisdom, for God has chosen the best possible world. It is not arbitrary, as with Descartes, nor necessary as with Spinoza. There is the actual series of states of the world, a series which might have been other than it is. For this series might have been altered. God has chosen this particular series of states from among others. At the center of the constitution of things is an act of purpose on the part of God, and this purpose is embedded in the substances which make up the world.

What makes up this "best possible world" which is chosen by God, not arbitrarily, as with Descartes, but by wisdom? For God's choice cannot be arbitrary. He does not act without reason.

On the one hand, we have God and the eternal truths; on the other, all those possibles which might be formed according to the eternal truths, among which is this particular and actual world. Each such possible world is a real possibility, specific and determined. But only one is chosen. Once again our mathematical model can be of help. That possible is chosen which has a "maximum of determination". Leibniz says,

> Suppose that from one given point a second point has to be reached, without any further condition to determine the path: the easiest or the shortest way will be taken.[17]

From one point to another, any number of paths could be taken. Each such path is specific, determined in the sense that its equation could be written. But only one such path would be *uniquely* determined, thus have a "maximum of determination", that is, there will be an infinite number of curved lines which may be drawn between the two points, but only one straight line. Thus, the "locus of all the points which are uniquely determined by their situation in relation to the two given points" is a straight line.[18]

It is the fact of a most determinate among all the possible worlds which provides a "reason" for creation. That, among all the sets of possible equations, given those general conditions called eternal truths, there is a most determinate, a necessary (but not sufficient) condition for an existing world. Were there no such most determinate world, God would not choose.[19] But the concept of itself cannot be the derivation of the existent apart from the choice of God. Leibniz never rules out the possibility that God will choose not to create at all. The world exists because of the will of God, not without reason.

Why does the world exist? Because God willed it. Why did God will it? Because this most determinate of all possible worlds *could* exist. The reason for God's choice is to be found in the reasonableness of this world, but it is a reason which "inclines but does not necessitate".

Leibniz asks the question which has become *the* cosmological question, the mind boggling one, "why is there something and not nothing?" The actual sentence is "However far you turn back to antecedent states, you will never discover in any or all of these states the full reason why there is a world rather than nothing, nor why it is such as it is." This question of Leibniz is one raised for the cosmos as Leibniz would describe it. And our thesis is that that cosmos, described by Leibniz as the "series of states of the world", is understood by him in terms of the "law of the series" discovered by him in the calculus. The cosmological question raised by Leibniz relates to the cosmos as he understands it. Would Leibniz have raised this question for *any* series of states? Would he have raised it for the series of states as described by Descartes? Would Leibniz look for a "sufficient reason" for the existence of the world if the world were simply extension and motion? The question is, of course, academic, since Leibniz states his question for the world as he knows and sees it: not for *any* world but for *this* world.

Note that the question is not only "why is there a world and not nothing" but "and why is it such as it is?" The cosmological question is here related to the teleological, as indeed it must be, for to ask "why is the world such as it is", one must ask "in comparison to what?" and thus "why is it as it is and not otherwise?" Leibniz here raises his cosmological question in the same way that Aquinas does: against the background of a specific cosmology, asking "why this cosmos and not another?" The cosmological question for both thinkers is intimately related to how the cosmos is. For Leibniz, the question "why is there a world and not nothing?" is framed for a world he has seen to be other than as Descartes thought it to be. So the question is "why *this* world, this world of intricate curvings and shadowed possibilities, and not nothing?" The question can be raised because of the nature of the world. As reason has been shown to be successful in explaining motion, so it can be assumed that it will be successful in explaining the world of which that motion is a part. And this particular world, a world which has rationality written into its substance, can be seen to be but one of many possible worlds. Things might have been different than they are. There might have been a world in which Caesar never crossed the Rubicon. There might have been a world in which the set of equations which go to make it up were radically simplified with just one such equation. Indeed there might have been a null set — no equations at all — no world at all.

Or, to set the same question in the same form in which Aquinas set it, Leibniz here asks, "why this world, so highly, indeed magnificently ordered, why is it so rational when it might have been so disordered? Why is it, not as Descartes thinks it is, but as I have shown it to be? Why is it that its motion can be expressed with finality as part of it and not simply as a random shuffling of parts? The answer must be that it is as it is and not otherwise because of the nature of God, who acts not by arbitrary whim but with wisdom and for the best."

So with Leibniz as with Aquinas, we find that the cosmological argument is intimately related to that cosmos from which they argue.

NOTES

1. C. I. Gerhardt, *Die Mathematischen Schriften Von G. W. Leibniz* (7 vols., Berlin, 1849–63), V. 301.
2. *Ibid.* V. 302.
3. C. I. Gerhardt, *Die Philosophischen Schriften Von G. W. Leibniz* (7 vols., Berlin, 1875–1890). II, 262–65. Also in L. E. Loemker, *Philosophical Papers and Letters* (Chicago, 1956), p. 533.
4. Gerhardt, *op. cit.*, II, 261. Loemker, *op. cit.*, II, 868.
5. *Ibid.*
6. G. P., III, 68.
7. *Discourse on Metaphysics*, Loemker, *op. cit.*, 485.
8. G. P., VII, 305. Also in P. Schrecker, *Monadology and Other Philosophical Essays* (N.Y., 1965), 89.
9. G. P., VII, 302. Schrecker, *op. cit.*, 85.
10. G. P., VII, 303. Schrecker, 86.
11. G. P., 232. Also in E. M. Huggard, *Theodicy* (London, 1952), 249.
12. Loemker, 514.
13. G. P., II, 49.
14. G. P., IV, 568. See also G. P., V., 140, 136.
15. Loemker, 413. Also in L. Couturat, *Opuscules et Fragments Inédit de Leibniz* (Paris, 1903), 513.

Chapter 7

CREATION, SCIENTIFIC EXPLANATION AND THE COSMOLOGICAL ARGUMENT

We now have enough background material in our study of two great philosopher-theologians to try to draw some conclusions related to our current situation. For, once again, we have a standard cosmology. The model which holds that we live in an expanding universe, which took its origin some 15 billion years ago in a primal explosion, is well established, has indeed become the "standard" model. As Aquinas framed his concept of creation against the background of an Aristotelian and Leibniz framed his against the background of a revised Cartesian, so we today when we speak of creation must speak against the background of this standard model.

With Aquinas, as with Leibniz, it is a model provided by the science of the day. The astronomers and physicists have said "this seems to be the way things work. This seems to be the way things have come to be." It is a cosmology which they present to us, namely that widest system of explanation possible by which to account for the whole of physical things. "This is its size; this its nature; this its age, according to the best understanding we now have." And how is the person of faith to respond to this?

The response, of course, is "God created it. Whatever it is you have come up with in your science as the whole of things, why that is what I mean when I say that God created. You are simply giving instantiation to my belief." And thus put, there can be no conflict between religion and science. As Aquinas would say of Aristotle's cosmos that it is derived from the will of God, so the person of faith would say that the cosmos, as delineated by science, is thus created. The constitution of things, the way things work, is all derived from science: the ultimate explanation of things is derived from faith. The statement of faith "God created heaven and earth" needs some exemplification, something to be pointed to, whether it is the earth and dome of Babylonia or the curved space–time universe of modern cosmology, something of which it is said "there, *that* is what I mean. God made all that." On the other hand, goes the argument, the scientist has no ultimate explanation of things. When once he has put forward the universe, he has reached the boundaries of science. He can explain no more, cannot explain the universe which he has delineated. And here faith steps in and says "the explanation which you cannot give is given in the act of faith in a God who creates".

Thus, both science and faith act in their own and proper spheres. Faith says nothing about the age or nature of things, and science says nothing about the ultimate origin of things, and they can thus live in peace together. Faith well knows that the science may change, may propose another and different model of things, and is content with this and somewhat indifferent to it, saying "no matter what you decide is the way things are, we will simply say that they are as they are because they are created by God. We have learned by experience that science will produce different models of things and that we will learn to live with that model. With some we have an easier time than with others. Aristotle's world was found to be very congenial, Descartes' rather forbidding, the modern cosmos, with its postulated first moment and boundedness, seems again congenial. But, while we have learned to live with various models of things, we have also learned never to marry one of them. Models may change but the faith endures, and part of that faith is the simple statement that 'God created heaven and earth'. You go ahead describing heaven and earth as best you can. We will continue to assert that they were created by God."

Now this mutual understanding is salutary. But there are some problems. They are problems raised with the scientific revolution of Galileo and Descartes. The basic problem is that science is in the explanation business. Its job is to explain more and more. And, as it does so, the question rises, is that explanation complete explanation? Granted that the world is as it claims it to be; is there in fact anything else? The statement of faith is indeed that all that is can be said to be created by God. And such a statement can indeed always be made. But is there any need for such further explanation? Is there any reason to make it? More and more things come within the explanation of science. Territory thought sacred has been invaded: miracles, religious experience, origins. Again and again, theology has seen things which it thought explainable only by theological categories invaded and claimed by science: design explained by evolution, religious experience by Freud, origins by chemistry. And, as this happens, the question must rise as to whether or not theological explanation of any kind is not, in fact, superfluous. Is not the universe, as science describes it, sufficient explanation of itself? Admittedly, no one theory can do all of this, while the theories themselves are often suspect and give place to other theories. Yet the question remains, as it did for early modern science with Descartes. Once given the scientific ways of things, the "laws of motion" as sufficiently and subtly understood, cannot science hope someday to explain everything? Will not science someday adequately explain religious experience itself, miracles, all questions of origins? And would it not then be best for religion to quit the field of explanation?

One way in which faith might deal with the increasing explanatory power of science is to say that science and theology explain the same thing but at different levels. A simple case in point is the crossing of the Red Sea by the Israelites. Could not the "miracle" of the crossing be explained both in a scientific way, by the wind pushing the water back, and in a theological way by the act of God in delivering his people? Thus, the theologian would be able to admit all that science would claim but at the end be able to say "yet all this is so because God has made his world so that it would be so".

Or, take the origin of life. To date, there has been no adequate way to explain the first replicating organism. Darwinian categories are valid only for replicating organisms, and thus cannot be used to explain those same organisms. Yet to say that no such explanation will suffice is dangerous, for there have been too many times when such explanations have, in fact, been forthcoming and one might very well be ready for publication that would adequately explain how, given the chemical background, such a replicating organism could arise. But, the theologian will say, even if that is possible, it is possible only because the chemicals are as they are, and for that, God must be the explanation.

Now there is an appealing austerity and simplicity about such a view of things. The difficulty with it is that of the camel in the tent. When once the camel is in the tent and there is no part of the tent which is not camel, is there really any place for the camel driver? Or, if the entire first floor can be explained by science, is there any need for a second floor? Especially if you have everything you need downstairs.

To point up the problem, take a look at the great Christian miracle of the resurrection. Suppose science came up with a credible explanation of the resurrection, say in terms of suspended animation and subsequent transformation of physical state: an explanation that would conform to the New Testament story, and then showed that, given a set of circumstances, such a resurrection would occur to anyone else as well. This is far fetched but may clarify the problem. Would we be willing to say that in such a case the theological explanation would fit this as it might fit the crossing of the Red Sea? And, if so, what would this do to the Christian faith?

Now you could, it is true, using the analogy of a two-story housc, retire to the second story, leaving the resurrection to its scientific explanation on the first storey while retaining an alternate theological explanation for the second storey. But the question would surely be raised as to whether or not that alternate view were anything but an extraneous alternate no longer needed by those who had managed to reduce this mystery to manageable terms.

As you retire from the first storey, you no longer, it is true, run the risk of being shown wrong, but do so at the risk of being extraneous. As theological explanation retreats again and again to the second storey, and the question "but why this world and not no world" raises no other question for science, it becomes more and more the second storey grandfather who descends less often to the family's daily area, becoming more remote and more dispensable. It is not simply a question of faith. One may indeed have faith that the second storey exists, but if there is less evidence of it and less use of it, one ends by living on the first floor and having the second only as an addendum, now and again remembered, but unnecessary to daily life. Thus, theology purchases its ability not to be wrong by its equal ability to be forgotten.

We need to look further at the great question "why something and not nothing?" and see if the cosmological argument it poses is valid apart from science. For only if so, can we truly retire to the second storey. Can the cosmological argument be shown to be an argument for God the Creator and be shown to be independent of any particular cosmology and any particular scientific explanation? We have claimed that the use by Aquinas and Leibniz of the argument was on the background of the cosmology of their day. The question is, can that argument be used independently of that or any science? Can we ask of anything, no matter what it is, "why does this exist instead of nothing?" And can we, as a result of that argument, postulate a God who is the reason for the existence of that thing?

It is my contention that we cannot. To support that contention, let me point out what it is we mean by the concept of God as "Creator of heaven and earth". It is meant that God here creates *ex nihilo*, that is, as framed by Aquinas, that there was nothing and then there was something. This is what we mean by creation. This is also what we mean by God, for the very notion of God as framed for faith is that of one who thus creates. A fully religious concept of God is that of one upon whom things depend for their very being, apart from whom they would not exist.

But this concept from its very nature is one of which we can have no experience. We never have any experience of an event in which something is not, in such an absolute sense, and then is. If, for example, we were to ask "why does the tree exist?", we would give a history of the tree, a chemical analysis of the tree, the personal preference someone had for the tree in planting it. And, in each case, we would be saying that there is a tree in this spot instead of a bush, or air, or the chemicals of which it is composed. Always we would be saying that there is a tree here when there might be something else. We might say that someone "made" the tree in the sense of producing a new variety of tree. But in no case would we expect to know why there is a tree here and not *nothing at all*. Yet that is what we mean by creation *ex nihilo*, namely, that the tree is dependent for its very being upon God, radically dependent, as is everything else, as is the universe itself. Something coming to be *ex nihilo* is as much beyond our experience as is something ceasing to be *intra nihilo*. The argument used against the cosmological argument as usually conceived is the argument against any demonstration of God as Creator, namely, that never in our experience do things either cease to be or come to be in this radical sense. The concept of creation *ex nihilo* comes not from our experience but from faith.

To answer the question "why is there something and not nothing?", you need to invoke the principle of sufficient reason as stated by Leibniz. The principle is that everything which exists must have a reason for its existence. And by this, Leibniz means not merely that you should be able to explain it in terms of preceding states, but that you should be able to explain the whole "series of states". You must be able to find a "sufficient reason" for things, something which itself requires no reason, something which exists "of necessity", which cannot not be. Of each thing you can say that it "might not exist", that it is contingent. But a full explanation must lead you, not simply to another contingent being, but to a necessary being. There must be a sufficient reason for the existence of things. God is such a sufficient reason because you need not ask of God *why* he exists. He exists of necessity, cannot not exist, and thus we have the explanation of things which we need. Of each individual thing, of the cosmos as a whole, we are able to say "it is contingent; it might not exist." Only of God can we say that he cannot not exist.

But this is to assert exactly what the concept of creation asserts: namely, that all things depend for their very being upon God, are of themselves dependent and contingent, might not exist. That God is necessary being and all others are contingent beings. And why should we accept the principle of sufficient reason on any other grounds than we accept the concept of creation, namely by faith? The principle of sufficient reason is stated by Leibniz on the background of the Christian doctrine of creation and asserts the same thing. And both, I would claim, while reasonable and compatible with ordinary scientific explanation, are not needed for that explanation or derived from it.

Why should we accept the principle of sufficient reason? There seems no inevitable reason to do so any more than there is inevitable reason to accept the concept of creation *ex nihilo*. And it is here that all efforts to state the cosmological argument, apart from any particular cosmology, must fail.

It is said, for example, that even had there been from eternity one atom, we should still want to know "what is the reason for this atom", since there might be no atom and thus nothing at all.[1] But is this true? Why should not the atom exist simply because it exists? We have no experience of "nothing at all" or of anything coming from or disappearing into "nothing at all". Quite clearly, to ask of this atom "why does it exist?" is to assume of it that it is a dependent thing, contingent upon some necessary being for its existence. And this is the same as to assert that the atom is a created thing called into being *ex nihilo*.

Here is the weakness of the argument given by Rowe in his *The Cosmological Argument*. Rowe revamps the argument of Clarke, itself a revamping of Leibniz's argument. He shows that some counter-arguments to the cosmological proof are themselves invalid. In particular, and in opposition to Hepburn, he shows that even if you have shown why each item in a set is there, you have not thereby shown why the set itself is there. The set of existents, argues Rowe, does raise a legitimate question as to why there is such a set. He shows that if this set were all "dependent", i.e., what they are because of something else, then there would be no answer to the question "why this set and not no set?" Using what he says is a weak form of the principle of sufficient reason, he says that there must be some necessary being as a member of this set of existents, and this necessary existent answers the question "why this set and not no set?".

The trouble here is that Rowe has failed to distinguish two strands to the question "why is there something and not nothing?". The one strand is similar to Aristotle's argument for the prime mover, namely, "why does this set continue to exist? Why has it not ceased existing?" And to that question, a necessary being is adequate. But this necessary being need not be God. It might, as Mackie argues in his *The Miracle of Theism*, be an eternal stock of matter.[3] It might be an eternal atom. It might, Aquinas would say, be an angel or a soul. The argument is Aristotle's for a prime mover, Aquinas' for God as the guarantor of continuing existence.

The other strand, here for Rowe the strong form of the principle of sufficient reason, asks not "why does this continue to exist?", but asks "why does it exist *at all*?" Even if it had always existed and never had a beginning, even if it were an eternal atom, we would still ask (tutored as was Aquinas by the Christian faith) "why has it come into being?" Not "why does it continue as it is?", but "why is it at all?" And the answer to *this* question, which takes us right beyond experience, is that of God the Creator. Here God is not only the one who guarantees the continued existence of things (as might an eternal stock of matter), but the one on whom things depend for their being.

Rowe correctly states that the question "why does the world have the members it does rather than none at all?" is a meaningful question. But it is a question which can be answered at two levels. The one is "why does the world with the members it has continue in existence?" The other is "what caused the world with the members it has to come into existence in the first place?" To the first question, the answer might be that there is a being which is permanent and thus ensures the continuing existence of the world. But that permanent thing might be an eternal atom, star or angel. At this level, the question is meaningful and seems to call for a definite answer. But of the second question, there seems no such clear-cut answer. The answer to the second question assumes the principle of sufficient reason in its equivalence to the concept of creation.

To say "why is there this set instead of no set?" means, by the "reason why", something more than one or a set of permanent things, something that explains not simply their continuation but their origination. To ask for a being from which this set has *derived* its being is to ask for God the Creator. But this is not a step which we seem required to make. When once we see that the set continues to exist because of some permanent being which cannot cease to be, we have come as far as reason can take us. To ask further, "but why this set and not nothing at all?", is to invoke the principle of sufficient reason farther than reason would require us to take it, is to invoke the concept of creation, a concept that is not of reason but of faith.

If the cosmological argument cannot be well framed apart from the cosmos from which it begins: if there is no valid argument in abstraction from any given scientific view: then to ask only "why is there something rather than nothing?" loses much of its force. But, in fact, the argument gains its force, not by its abstraction, but by the concrete situation in which it is framed. There is a way to put the cosmological question with such force that the theological answer again becomes relevant. It is the way of Aquinas and of Leibniz. Each of them spoke out of a science of their time, a cosmology of their time. Their argument for God was never in isolation from that science. For Aquinas, the science and cosmos were that of Aristotle. He asks, "is it not obvious that such a cosmos, so beautiful, so ordered, would, if only on its own, in time return to that from which Lucretius said it had come — atoms and the void?[4] And that, once dissolved into that infinite chaos, it would never return? And is it not equally obvious that there must be some being and/or beings which, in their own enduring, uphold the fabric of things?" That being or beings need not be God the Creator: but it could be.

Leibniz too argues from the science of his time, a science in which he had a part. He says "is it not rational to believe that so rational a world was itself framed by reason?" No doubt he would assert that any and all things, whether rational or not, whether well ordered or not, must have a sufficient reason for their existence. But *in fact* the world of which he asserts God as necessary being is a world of rationality, and it is that very world which gives the argument its force. We might, it is true, ask of a world of pure gas in random motion "why does it exist and not nothing?", but that question would seem weak and the answer, "perhaps it just exists because it exists", easily given. But when that world is one of intricate and subtle weavings of lines, beautifully and rationally calculated, the question takes on great weight, and the Christian answer takes on authority.

In short, what you assert about the world makes it easier or harder to get from that world to God. This is not to wed theology to a particular cosmology: on that way lies disaster. It is simply to say that some pictures of things raise the cosmological question "why something rather than nothing?" with special urgency, and do so with the question "why *this* something, this particular cosmos as it has been described by science?"

Now this is obviously not a "proof" in the usual sense of the word. For any argument that depends on the historical circumstances of its time is obviously limited to that time. But this, which would be a weakness for the ordinary argument, now turns out to be a strength. The proof of God as Creator of heaven and earth is beyond us. There is a cosmological proof, but it turns out to be simply of an enduring "somewhat", not of God the Creator. If God the Creator is in any way to be known from his creation, it is not from a proof, but from the sheer weight of the question, "why is there *this* something rather than nothing?"

Now the "weight" of the question depends on the science of the day. The cosmological argument, if abstracted from that given science, leads us only to a permanent "somewhat". But, if embedded in that science, and if that science is a highly articulated and structured one, which is part of the whole world view, then the question becomes weighty indeed. It is a question which science need not answer: indeed, cannot answer, for it leads beyond its domain. Yet it is a question which will not down. When contemporary cosmology speaks to us of a "big bang", and of a finite yet unbounded universe which contains all the space and time there is, we *must* ask "why?" And there seem but two answers which can be given. They are the answers we give to an inquisitive child who will not be satisfied with any answer: either "God made it that way" or "because that's just the way it is". The two answers give two world views. One is of a one-story universe; the other, of a two-story. Which answer are we to give? And what reasons have we for giving the answer that we do?

The reply must be that we make our choice by factors and arguments which are many and varied, some of them ones of which we are hardly conscious. All kinds of hunches, intuitions, reasons play their part. We choose between two views which are both possible views, both intellectually respectable views. To say that the world exists because God created it is not to vitiate a single scientific principle. The world remains as science says that it is. Theology here takes the picture that science hands it, alters it not an iota, and simply says, "but all this is a created thing. And by creation, I mean creation *ex nihilo*, a making beyond any scientific category and for which experience provides no guide."

The situation is not unlike that posed in accepting a scientific theory. A theory is rarely proven or disproven. It ceases to apply. Somehow it no longer seems to work and vanishes from the scene, and no one can really say why. There were difficulties with it. A rival theory seemed simpler. So epicycles, or phlogiston, or the ether vanish before heliocentrism and oxygen and relativity.

Yet the choice cannot be arbitrary. There must be some reasons for the choice that is made. Let me point out some for choosing a two-storey world, a world with God in it, a God who is Maker of heaven and earth. This is, if you will, the landing between the stories that indicates that there is indeed a second storey.

There seem to be two in contemporary cosmology: one is that of the beginning point postulated in the standard model; the other is that of the arbitrary constants of that model. There will obviously be many others, drawn from the moral life, drawn from the fact of consciousness, owing their strength and weakness both to the science of the time and the persuasions of the one asserting them. But these two will do to illustrate the point.

If, as the standard model now seems to assert, there is a beginning point, behind which our knowledge cannot seem to go, so the equations all have a first moment, then the question inevitably arises, "where then does the cosmos come from? What is before this?" Aquinas himself saw the force of this argument. He realised that the concept of creation was by no means dependent on such a first moment, but he said that such a first moment was an "aid to the imagination". So, in thinking about creation, it is of help to think of just such a first moment. It is of help to us as well in grasping the meaning of the concept of creation. "The cosmos comes to be and has a beginning in time because God made it" is both an instantiation of our concept of creation and an argument for it. Certainly it is not an overwhelming argument. A first moment is no proof of a Creator. But that it can be evidence for a Creator, certainly weights the question "why this something with its first moment?" inevitably in the direction of the answer of faith.

A second argument is from the arbitrary constants which are at the base of things. Let me quote from a more or less standard text in cosmology, that by Edward R. Harrison.[5]

> So far we have spoken as if just one universe existed. In this section we suppose that not one, but many, physical universes exist and that each is self-contained and unaffected by all the rest. In each physical universe the fundamental constants of nature have different values. By constants of nature we mean permanent things, so far unexplained, such as the speed of light, c, the gravitational constant, G, Planck's constant h, the electric charge, e, of the proton and the electron, and the masses of the subatomic particles. We have hence an array, or ensemble of universes that covers all values of the constants of nature. Each universe is a workshop in which we can study what happens when the constants are assigned specific values different from those in our own universe.
>
> We first notice that alterations in the known values of c, h and e cause huge changes in the structure of atoms and atomic nuclei. Even when the changes are only slight, most nuclei are unstable and cannot exist. The majority of universes in the ensemble contain little more than hydrogen. . . . We also find that slight changes in the values of c, G, h, e and the masses of subatomic particles cause huge changes in the structure and evolution of stars. The majority of universes will not contain any stars, and in the few that do, the stars either are non-luminous or are so luminous that their lifetimes are too short for biological evolution.

> Life forms . . . depend on a habitable environment, such as a planet warmed by a long-lived star, in which it can originate and evolve. . . . Our universe would not exist if the constants of nature had different values.
>
> This can be interpreted in two ways. The first is that the ensemble is real and only our universe and perhaps others closely similar contain living creatures. . . . The second interpretation is that a Creator has designed our finely tuned universe specifically for the containment of life.

Now how should a person of faith respond to this? Should he say, "Well that's just what the scientist says now. He will change his mind in time and explain it some other way. My faith doesn't depend on what science says anyway."?

There is another, perhaps subtler, argument for creation in our modern cosmology, and that is the extent to which rationality has shaped it. Cosmology of our century goes back to Einstein and the general theory of relativity. It is with that theory, and its use of Riemannian space, that it is possible to talk of the cosmos again in a comprehensible and rational way. For, once again, as it was for Aristotle, though in a much subtler and sophisticated way, the cosmos becomes discriminable, something the mind can delineate. It is so because it is once again finite, though unbounded. A light ray could presumably travel through the universe and return to its starting point without ever encountering a boundary, a "thus far shall you go and no further". Yet it would have travelled through all the space there was, for space itself as space–time is curved back upon itself. There is no space "outside" this finite universe as there is no time "before" it. Space and time are no longer absolutes, but part of the universe. This means that it is legitimate to talk of "all there is" and see it as a discriminable and circumscribed whole. "This is what I mean when I say God created heaven and earth", the theologian can say: "this whole described by cosmology."

But there is something even more. For this whole is penetrable by reason, a reason based on a mathematics of the space–time continuum which in itself predicts phenomena unexplainable by Newtonian mechanics. Newton's system is essentially empirical. Newton never knew the cause of gravitation. There is a "somewhat" which makes the planets revolve and the apple to fall, a "given" which can be expressed in the Newtonian mathematics. But that which is a simple "given" for Newton is explained by Einstein. Einstein is not content to know that the Newtonian laws of motion are as they are: he seeks to explain why they are as they are. Thus, for example, the identity of inertial and gravitational mass are pure "givens" for Newton. It is a physical miracle that the mass which is pushed and the mass which responds to gravitation are one and the same. But for Einstein they are one and the same because both are part of the space–time manifold in which the mass moves. It is seen that, given that space–time manifold, they *must* be the same. Thus, in a triumph of rational reasoning, Einstein is able to explain this identity, as well as to predict the bending of light rays near a star, the advance of the perihelion of Mercury, the slowing of time near the sun. There is a sense in which Einstein shows why things move as they do in the same way in which Aristotle sought to show why things move as they do. For both there is a "natural" motion, only much more subtly so for Einstein. Things move as they do because of the nature of space–time, not because of a mysterious force of gravity or because they are pushed or pulled.

Now the fact that the universe can be known by reason is not itself an argument for God or for creation. But it can be such an argument when that rationality points us to a discriminable universe and when that universe is as it is because of what seem to be arbitrary constants. Were a unified field theory possible which would include within it atomic forces, could a mathematical theory explain why those constants are as they are and cannot be other than they are, then the force of the argument would be removed. We could then say, as Spinoza would say, that things are as they are because they cannot be other than they are. But that is not the case. There is place not only for the brilliant deductive reasoning of an Einstein and a world knowable by rationality but also for the patient working of the laboratory experimenter to find how things in fact are. There is both rationality and contingency: a universe knowable by reason and also with a "givenness" about it which reason cannot subdue. Such a universe points to a creative intelligence who creates, not out of necessity, but out of will.

The objection, of course, will be raised, "but these things depend on the science of the day. Is belief in God, then, also dependent on what cosmology may teach in any given decade? Shall I believe today and not believe tomorrow according to the latest cosmological discovery?" And the answer must be "Who knows what finally decides our choice for one view of things compared to another? It is not simply one discovery or another. We are dealing with a whole view of things in their entirety. All kinds of factors make up our final decision. What cosmology has to say may well be one of those. It is doubtful that that alone would ever be the deciding factor, and certainly never the only factor. Logic and sermons, as Whitman observed, rarely convince."

There is another way. It is to look again at the cosmological argument and the teleological argument and to use them, not as proofs for the existence of God, but as evidence for that existence. The cosmological argument, I have maintained, winds up giving us simply a permanent "somewhat", a necessary being in the sense of one which endures and in its enduring ensures the continued existence of things. But if God exists, then this is an explanation of that enduring. The same can be said of the teleological argument, which asks "why is there this something and not something else?" This is the argument that is related to the science of the day, for it must always begin with a particular something. It has been my argument that both Aquinas and Leibniz shape their cosmological arguments with this teleological slant. It is immune to the questions that can be raised about the "nothing" of the cosmological argument, a "nothing" of which we have no experience. But we can understand that our world is ordered, can understand that it might not be ordered. We know what order is, can ask "why order and not disorder: why this articulated world when all might very well be simply a gas?" And the fact of order seems to call for one who designs. The objection to this is that the most that it shows is that God is Designer, not Creator. But it can be used as evidence for God as Creator: surely if God is Creator he is also Designer. There may be other explanations for that which endures and that which shows design, but belief in God as Creator is certainly a powerful explanation. And the more order is discovered, the more highly structured the cosmos is found to be, the more and more weighty is that question which must come: Why? Why this order which science increasingly uncovers? And, while the answer is still tenable, "the cosmos is simply because it is"; the answer of the faith, that it is because God has created it to be so, takes on more and more authority.

There is, of course, a risk in this. For, if science becomes more and more able to explain those things which theology has explained in the past, a one-storey world becomes more and more to be the likely world. If miracles, including the resurrection, if religious experience, if the fact of consciousness and conscience, become more and more explainable by science as a comprehensive and purely secular view of things, then the view that all of this is created by God to be as it is, becomes, while intellectually tenable, less and less a working part of intellectual consciousness. It thus behoves us to pay attention to what is going on in the lower storey, and to make it a point to descend as often as possible to remind the family that there is more to things than they had dreamed, and to invite them to learn to live with a two-storey consciousness.

There is, in short, no one royal staircase to God, fixed grandly and forever despite the shifting first floor floor-plan. The reasons for belief in God, for making that great statement "I believe in God the Father Almighty, Maker of heaven and earth", vary with the science of the day and the accidents of the day. Faith, of course, the faith that "casts itself on God for life and for death", remains, with or without its reasons. But the evidence shifts from one era to the next, and, at any given time, may be weaker or stronger, depending on the science of the day. And that is the risk of faith.

NOTES

1. Richard Taylor, "Metaphysics and God" in *The Cosmological Arguments*, D. Burrill (ed.), (N.Y., 1967).
2. W. L. Rowe, *The Cosmological Argument* (Princeton and London, 1975).
3. J. L. Mackie, *The Miracle of Theism* (Oxford, 1982).
4. *On the Nature of Things*.
5. E. R. Harrison, *The Universe* (Cambridge, 1981).

BIBLIOGRAPHY

Chapter 1

For a non-realist view of science, see Bas Von Fraassen, *The Scientific Image* (Oxford, 1980). Also Larry Lauden, *Progress and Its Problems: Toward a Theory of Scientific Growth* (Univ. of California, 1977). An Instrumental view is found in Ernest Nagel, *The Structure of Science* (Harcourt, 1961). A résumé of the views is in Ian Barbour, *Issues in Science and Religion* (Prentice, 1966), 162ff.

Non-cognitive views of religion are given and criticised in Barbour, 238ff. An emphasis on the act of faith apart from its content in Tillich is found in his *Dynamics of Faith* (Harper, 1957).

For Genesis: John Skinner, *Genesis: The International Critical Commentary* (Edinburgh, 1930); E. A. Speiser, *Genesis: The Anchor Bible* (New York, 1964).

On early Christian cosmology, see A. E. White, *A History of the Warfare of Science with Theology in Christendom* (New York, 1901). The fullest treatment of the interpretation of Genesis 1 is in D. O. Zöckler, *Die Geschichte der Beziehungen Zwischen Naturwissenschaften mit besonderer Rücksicht auf Schöpsungs-Geschichte* (Gütersloh, 1977).

Chapter 2

I have used interpretations found in Marjorie Grene, *A Portrait of Aristotle* (Chicago, 1963) and J. H. Randall, Jr., *Aristotle* (Columbia, 1960); also Friedrich Solmsen, *Aristotle's System of the Physical World* (Ithaca, 1960).

Chapter 3

A full discussion of the five proofs is in Kenny, *The Five Ways*. The standard interpretation of Aquinas' third way is in Frederick Copleston, *A History of Philosophy* (Image, 1961), vol. 3, Part 1. Trenchant criticism of the traditional argument is in J. L. Mackie, *The Miracle of Theism* (Oxford, 1982).

Chapter 4

For Aquinas on creation, see especially A. D. Sertillanges, *L'Idée de Creation* (Paris, 1945). Also Garrigou-Lagrange, *The Trinity and God the Creator* (St. Louis, 1942).

Chapter 5

For the Copernican Revolution, most helpful is Thomas S. Kuhn, *The Copernican Revolution* (Harvard, 1957).

For Galileo, see Ludovico Geymonat, *Galileo Galilei* (New York, 1965) and G. de Santillana, *The Crime of Galileo* (Chicago, 1955).

For Descartes and the mechanical philosophy, see the pioneering work of A. E. Burtt, *The Metaphysical Foundations of Modern Physical Science* (New York, 1926).

Also the studies by A. Koyré, *From the Closed World to the Infinite Universe* (New York, 1959), *Études Galiléennes* (Paris, 1939), and *Entrétiens sur Descartes* (New York and Paris, 1944).

On eternal ideas, see "Creation of Eternal Truths in Descartes' System", in *Descartes: A Collection of Critical Essays*, ed. W. Doney (New York, 1967).

For Descartes' Physics, see Paul Mouy, *Le Développement de la Physique Cartésienne, 1646–1712* (Paris, 1934).

Chapter 6

Sources for Leibniz are in *Sämtliche Schriften und Briefe*, ed. Prussian Academy of Sciences (Darmstadt and Leipzig, 1923–); C. I. Gerhardt, *Die Philosophischen Schriften von G. W. Leibniz* (7 vols., Berlin, 1875–90); and C. W. Gerhardt, *Die Mathematischen Schriften von G. W. Leibniz* (7 vols., Berlin, 1849–63).

Space and time are treated by H. G. Alexander, *The Leibniz Clarke Correspondence* (Manchester, 1958).

Basic interpretations of Leibniz can be found in N. Rescher, *The Philosophy of Leibniz* (Englewood, N.J., 1967) and Bertrand Russell, *Critical Exposition of the Philosophy of Leibniz* (2nd ed., London, 1937).

For Leibniz's dynamics, see Pierre Costabel, *Leibniz et la Dynamique* (Paris, 1960) and M. Gueroult, *Dynamique et Metaphysique Leibniziennes* (Paris, 1934); Max Jammer, *Concepts of Force* (Cambridge, Mass., 1957) and Paul Mouy (*op. cit.*).

Leibniz's mathematical discoveries are in C. B. Boyer, *The Concept of the Calculus* (New York, 1939), 187ff; see also Leon Brunschwieg, *Étapes de la Philosophie Mathématique* (Paris, 1912), 171ff.

Chapter 7

For an exposition of the "big bang" model of cosmology, there is Steven Weinberg's *The First Three Minutes* (New York, 1977).

The "anthropic principle" is to be found in B. J. Carr and M. J. Rees, "The Anthropic Principle and the Nature of the Physical World", *Nature* (April, 1979), p. 605 and in Sir Bernard Lovell, *In the Center of Immensities* (1976), pp. 119, 122ff. See also S. J. Jaki, *Cosmos and Creator* (Edinburgh, 1980).

On the cosmological argument, see Milton Munitz, *The Problem of Existence* (New York, 1965). See also W. L. Rowe's *The Cosmological Argument* (Princeton and London, 1975). The most recent study and one which also uses the traditional arguments for God as evidence rather than deductive proof is by Richard Swinburne, *The Existence of God* (Oxford, 1979). Trenchant criticism of the traditional argument is in J. L. Mackie, *The Miracle of Theism* (Oxford, 1982).

The rationality of Einstein's physics is expounded by Cornelius Lanczos in "Rationality and the Physical World" in *Boston Studies in the Philosophy of Science*, vol. 3 (1967).